T0257109

HEALTH CARE IN
BIRMINGHAM

The Birmingham Teaching Hospitals
1779–1939

In the middle of the eighteenth century, hospitals were unfamiliar institutions to the inhabitants of most English towns and cities. As early as the late nineteenth century, however, hospitals had become central to both the provision of health care and medical education in most large urban population centres. Drawing on hospital records, the publications of associated medical staff and a wealth of other local documents, *Health Care in Birmingham* carefully maps the evolution of nine voluntary hospitals, and their associated medical specialities in Birmingham, England over the century and a half before the introduction of the National Health Service, a period that witnessed significant social, economic and cultural change. From the emergence of the town's first General Hospital in 1779, the wealth of this key industrial centre in particular encouraged the development of a full range of medical institutions, including those established to treat afflictions of the bones and joints, eye, ear, teeth and skin, as well as ailments peculiar to women and children. Besides charting the local development of a wide range of specialist fields, *Health Care in Birmingham* firmly situates each hospital in its local and national contexts. Though greatly reorganised on the eve of the Second World War, these institutions influenced considerably the history and landscape of the city, and continue to do so today. This is the first time their history has been considered collectively in a single volume.

Jonathan Reinarz is Director of the Centre for
the History of Medicine at the University of Birmingham.

HEALTH CARE IN
BIRMINGHAM

The Birmingham Teaching Hospitals
1779–1939

Jonathan Reinarz

THE BOYDELL PRESS

First published 2009
The Boydell Press, Woodbridge

ISBN 978 1 84383 506 6

The Boydell Press is an imprint of Boydell & Brewer Ltd
PO Box 9, Woodbridge, Suffolk IP12 3DF, UK
and of Boydell & Brewer Inc.
668 Mount Hope Ave, Rochester, NY 14604, USA
website: www.boydellandbrewer.com

A CIP catalogue record for this book is available
from the British Library

The publisher has no responsibility for the continued existence
or accuracy of URLs for external or third-party internet websites
referred to in this book, and does not guarantee that any content
on such websites is, or will remain, accurate or appropriate.

This publication is printed on acid-free paper

Designed and typeset in Adobe Caslon Pro by
David Roberts, Pershore, Worcestershire

Printed in Great Britain by
CPI Antony Rowe, Chippenham and Eastbourne

Contents

Illustrations

All images except no. 1 reproduced courtesy of Birmingham Libraries and Archives.

Dedicated to the memory of
Joan Lane (1934–2001)
Teacher, mentor and friend

Foreword

THIS IS THE FIRST scholarly history of Birmingham's teaching hospitals and is the result of a three-year research fellowship held by Jonathan Reinarz. It is also the history of many hospitals that are no longer with us, and this book is therefore a tribute to these institutions that served the city and its population so magnificently in war and peace, going back to the end of the eighteenth century. Both of us have very fond memories from many years either as a medical student and doctor or a patient at the General Hospital.

The author was originally employed by the Centre for the History of Medicine at what was then the University's Medical School to research and write this volume. Although he has now, through the generosity of the Wellcome Trust, become a full-time Lecturer in the History of Medicine and is now the Centre's Director, he continues with this work and over the years we will see ever more about the medical history of the United Kingdom's Second City emanating from his pen. Owing to his efforts, the medical history of our city is firmly on the map.

The original research was funded by the Special Trustees of the Former United Birmingham Hospitals, now split into a number of charities associated with the various NHS Trusts and Foundation Trusts that are the successor bodies to the United Birmingham Hospitals. It is important that we acknowledge that this book would not have been possible without their support and those who formed the first Advisory Group, especially the late Dr Ben Davis and the late Dr Joan Lane.

We would also wish to warmly thank the Sir Arthur Thomson Charitable Trust who has made publication possible by a generous grant. The late Sir Arthur was, following his retirement as Vice Principal and Dean of Medicine and Dentistry of the University of Birmingham, an ardent sponsor of Birmingham medical history and we are sure he would have greatly enjoyed reading this book.

Professor Robert Allan
Professor Robert Arnott
January 2009

Acknowledgements

THE THREE-YEAR PERIOD during which the majority of the research for this book was conducted was funded by the University Hospitals Birmingham NHS Trust. I am very grateful to the trustees of this charitable body for the opportunity to research and write the history of Birmingham's first teaching hospitals. I particularly appreciate their decision to appoint a historian, and a foreign one at that, to this task. However, over the six years that were required to complete this project, I have become indebted to many more individuals and organisations.

From its outset, the project was managed by a dedicated steering committee. Its nine members regularly provided me with their support and expert guidance, not to mention valuable historical information. Chaired by Professor Robert Allan, this group of medical practitioners and historians included Robert Arnott, the late Dr Joan Lane, Dr Leonard Schwarz, Dr Len Smith, the late Professor Owen Wade, Stuart Wildman, Professor Alexander Williams and, in the project's final year, Dr Martin Gorsky. Rivalling even the most generous of nineteenth-century philanthropists, Leonard Schwarz additionally provided me with meals, accommodation and stimulating conversation more recently when on my regular visits to Birmingham. Robert Arnott, in turn, managed the final negotiations with publishers and helped eventually to get this story into print.

Over the project's six years, the steering committee also placed me in contact with a number of local health professionals. I am grateful to these individuals who formerly worked at a number of the teaching hospitals discussed in this volume. Many were interviewed in the first months of the project and their views helped shed valuable light on the development of Birmingham's voluntary institutions, and allowed me to orient myself in a new field of historical research. They included Professors Graham Ayliffe, J. M. Bishop, D. B. Brewer, B. L. Pentecost, Sir Geoffrey Slaney, Messrs Leon Abrams, Victor Brookes, Ronald Cohen, Ron Fletcher, James Inglis, Harold Lilly, G. D. Oates, Owen Wade, George T. Watts and Mrs Barbara Gaunt.

The majority of the records of the Birmingham hospitals have been catalogued and are maintained by the staff at the Birmingham City Centre Archives. It was here and in the Local Studies section of the Birmingham City Centre Library that I spent the majority of my time producing my own personal archive of data during the project's first stages. I am grateful to the archive's staff for aiding my research over innumerable visits to their collection. In particular, I am grateful to Sian Roberts for facilitating very healthy and useful communication between the archive, this project and the Centre for the History of Medicine at the University of Birmingham. I also wish to thank several members of the archive staff for assisting with my research since my arrival in Birmingham, including Richard Abbott, David Bishop, Rachel MacGregor, Corinna Rayner, Angela Skitt, Alison Smith and Fiona Tait.

Material relating to the medical school is found in the University's Special Collections, until recently under the watchful eye of Christine Penny. I am particularly grateful to its staff, including Phillippa Bassett, Anne Clarke, Martin Killeen and Helen Fisher. Many other reports and original publications were accessed at the Barnes Library and the library of the Birmingham Medical Institute. Finally, the collection at the Wellcome Trust Library in London, including the work of many individual historians of medicine, helped contextualise the story that emerged from these local repositories.

I am grateful to a number of historians, colleagues and medical professionals who spent many hours discussing various angles of the project with me, assisting with images, maps and tables and eventually reading my written drafts. In addition to the members of the project's steering committee, these include Jane Adams, Sam Alberti, Fran Badger, John Breuilly, Amanda Cadman, Tim Carter, Stephen Cherry, Pamela Dale, Malcolm Dick, Erika Dyck, Paul Elliott, Geraldine Goodman, Neil Handley, Anne Hardy, Clare Hickman, Matthew Hilton, Julia Hyland, Colin Jones, Peter Jones, Judith Lockhart, Bill Luckin, Anna Lyon, Hilary Marland, Rupert Millard, Alex Mold, Richard Moore, Dmitri Nepogodiev, Peter Nolan, John Pickstone, Noelle Plack, Alistair Ritch, Shelley Smith, Anne Spurgeon, Andrea Tanner, Carsten Timmermann, Helen Valier, Rachel Waterhouse, Ruth Watts, Andrew Williams and the late John Hutchinson. I am especially grateful to Henry Connor, Graham Mooney and Keir Waddington who read through the entire manuscript at least once when they had many more pressing deadlines pending. Though only there for the project's first year, Joan Lane's early advice and generous support provided this project with considerable guidance from the start and was an inspiration to the end. She also greatly aided my transition from a social to a medical historian in the years leading up to the project's outset.

I am also grateful to the staff of Boydell & Brewer, who agreed to publish this volume and improved its presentation along the way. In particular, I wish to thank Peter Sowden for his early help with negotiations and production, Michael Middeke for his assistance with the script, Catherine Larner for her guidance with images, Vanda Andrews and especially David

Roberts for their help with the proofs, not to mention their early feedback on the text.

Given my unusual living arrangements in the project's first two years, residing part of the year in Vancouver, Canada, between intensive periods of research in Birmingham, England, I am very appreciative of the support provided by friends and family during this period. In particular, I wish to thank Duncan Dee, Neil Killey, Reine Mihlta, Trevor Shew, Adrian Staehli, as well as my brother, Jerome, and parents, Monique and Frederik.

Finally, I am particularly indebted to Marsha Henry, who lived with the project since its conception, and Theo and Audrey, who took much less time to gestate, though come far closer to perfection than does this historical study.

List of Abbreviations

BCLA	Birmingham Central Library Archive
BCLLS	Birmingham Central Library Local Studies
BCU	Birmingham Cripples Union
BDHCL	Birmingham Dental Hospital, Cohen Library
BECC	British Empire Cancer Campaign
GMC	General Medical Council
LDS	Licentiate in Dental Surgery
ROSH	Royal Orthopaedic and Spinal Hospital
UBSC	University of Birmingham Special Collections

A History of the Birmingham Teaching Hospitals, 1779–1939

IN THE AUTUMN OF 1779, William Jones, a labourer from Lichfield working and residing in Birmingham, seriously wounded his hand as the result of an accident, apparently while working a lathe. Although Jones and many other labourers like him had without a doubt previously endured similar injuries at home, in the community and in the workplace, the twenty-nine-year-old's experiences of health care on this occasion would be very different from most of his local contemporaries. Thanks to the fore-sight and generosity of Samuel Baker, a turner, who lived and worked from premises on Birmingham's High Street and had recently become a sub-scriber to the General Hospital, Jones would not have to pay for his medi-cal treatment. In return for his donation to the newly constructed General Hospital, Samuel Baker was entitled to recommend several patients to the institution, one of which was his employee, Jones. This happened some days after the initial accident, presumably when the worker's wound showed signs of infection.

Once at hospital, Jones's hand was cleaned and bandaged by the char-ity's apothecary. Over the following days, he was carefully observed by Robert Ward, the surgeon who periodically visited the hospital to attend nine other surgical patients, three of whom were women, during the insti-tution's first week of admissions. During subsequent weeks, the labourer's hand underwent regular dressing and undressing, was doused with vari-ous medical preparations, treatment which was occasionally carried out by Ward's apprentice, who had also been granted access to the hospital's wards. Jones was discharged from hospital more than two months later on 11 December and remained an outpatient until 11 March 1780. Though pos-sibly a difficult case, whose hand experienced serious infection and perhaps the loss of several digits if not some movement, Jones was one of the hos-pital's first success stories. Along with all but one of the patients admitted on the same week in October, the labourer was eventually discharged as 'cured'.[1] In subsequent years, not all medical emergencies would conclude as successfully, but many more of Birmingham's inhabitants, not to men-tion pupils and practitioners, would be admitted to the wards of similar

---

[1] Birmingham Central Library Archives (BCLA), General Hospital (GH)/4/2/560.

medical institutions and begin to experience medicine and health care in a
hospital environment.

By this time, it would be fair to say that the town's inhabitants had
become somewhat accustomed to such dramatic forms of change. No
longer just an important regional market, Birmingham by the 1750s was
well on its way to becoming the manufacturing centre of the Midlands.
The town was divided into a lower industrial district, comprising workshops and warehouses, and an upper residential area, funded by the proceeds of local industry and made up of newly laid streets that converged in
handsome squares. To some visitors it resembled another London, albeit
in miniature.[2] Much of its wealth, the result of a concentration of metal
industries in hundreds of workshops, was derived from a particularly plen-
tiful supply of iron ore and coal located in nearby South Staffordshire and
East Worcestershire. Great supplies of both resources were transported to
Birmingham by way of a network of turnpike roads and transformed into a
wide range of consumer goods, from ornaments to armaments. From 1740,
industry further diversified with the manufacture of brass, metallurgical
trades having replaced agriculture as the most important sector of the local
economy. With the help of some rudimentary machinery, many of which
were invented by the town's inhabitants, and a limited number of engines,
as well as new methods of labour organisation, available raw materials were
transformed into buttons, buckles and, most famously, pins. Within a few
decades, Birmingham was described as 'the first manufacturing town in the
world'.[3]

While the markets for these consumer durables boomed, so too did
the local labour market, as Birmingham moved from being the fifth largest town in England in 1750 to the third largest less than three decades
later, with a population of 40,000; only London (775,000) and Bristol
(55,000) were larger.[4] Further growth was fuelled by the development of a
local financial system in the 1760s with the appearance of Taylor & Lloyd,
Birmingham's first bank, and the exploitation of foreign markets, both on
the Continent and in America. Despite the early existence of much distant trade, markets nearer to Birmingham were by no means more accessible or developed. For this very reason, a sizeable share of business profits went towards improving local means of communication. By the 1760s
this largely meant improving the region's 'despicable' roads, and building

---

[2] E. Hopkins, *The Rise of the Manufacturing Town: Birmingham and the Industrial
Revolution* (Stroud, 1998), 3.

[3] Hopkins, *Rise of the Manufacturing Town*, 26. For the famous case of pin-
making see A. Smith, *The Wealth of Nations* (London, 1987), 109–10.

[4] Ibid., 21.

an extensive canal network, first through the Black Country (1770), followed by routes to the east through Fazeley (1783), as well as Dudley and Stratford (1793) in order to reduce the transportation costs of bulky raw materials and manufactured goods.[5] Only by reducing the costs of production would Birmingham's industries retain the advantages its manufacturers had achieved. To many others, the strength of future trade required ensuring the health of the town's labouring population.

Medical provisions for those who happened to suffer industrial injuries or simply succumbed to the effects of poor diet and their new urban environment did not immediately appear to deteriorate with the rapid rates of growth the region experienced in these decades. Many would have been convinced of this following the pronouncement by Joseph Priestley, the chemist and discoverer of oxygen, that the air of the town was 'equally pure as any he had analysed'.[6] In general, Birmingham was perceived as a healthy town, and its labourers, kept in constant employment, purportedly enjoyed good health. Industry not only kept their bodies fit, but improved the quality of the food they ate and, as importantly, kept their minds employed.[7] In fact, to the town's privileged inhabitants, the health of Birmingham's workforce appeared to be threatened more by luxury than labour.[8]

Should health care services have been required, however, the town's industrial wealth and successful middle-class entrepreneurs had attracted many medical practitioners to the region. In 1767, there were at least twenty surgeons and three physicians serving the local population.[9] The elite of the local medical community, as elsewhere, were the university-trained physicians, who were not only acquainted with the sciences and literature of their profession, but possessed that amount of general knowledge which distinguished the scholar and gentleman. Those occupying the highest medical rank locally included the Shropshire-born and Edinburgh-educated William Withering, more often associated with his work on digitalis, and William Small, who obtained his education, or at least a recognised medical qualification, in Aberdeen. Though the size of medical practices varied, Small was earning in excess of £500 annually in the 1760s, while Withering's professional income exceeded £1,000 a decade later. Surgeons active in these years included Edmund Hector, a schoolfellow and intimate friend of Dr Samuel Johnson who settled in Birmingham in

---

[5] W. Hutton, *History of Birmingham* (Birmingham, 1781), 263.

[6] C. Pye, *A Description of Modern Birmingham* (Birmingham, 1819), 4.

[7] T. Tomlinson, *Medical Miscellany* (London, 1769), 37.

[8] Ibid., 89.

[9] J. Lane, *A Social History of Medicine* (London, 2001), 23.

1732, and Francis Parrott, who, according to another local surgeon, James Russell, never performed an operation in his life.[10] In contrast, Thomas Tomlinson provided his colleagues' apprentices with the first series of practical anatomy lectures in the town in 1767. Even a moderate share of practice, however, limited a surgeon's time for such educational ventures.[11] Besides training the odd apprentice, local surgeons established private practices, served local Poor Law authorities, and a vast network of local friendly societies and other mutual associations.[12] In fact, several thousands of the town's workers were members of friendly societies, the nearby Black Country having had the largest concentration of such clubs nationally.[13] Those labourers unable to join such insurance schemes generally relied on the support of friends and family, or occasionally on the charitable services of a medical practitioner prepared to surrender payment for medical services, at least to those individuals fortunate enough to share a common religious affiliation. Alternatively, from 1779, many more local cases, including William Jones, would enter the town's first voluntary hospital, the medical services of which were equally dependent on the charitable work of medical staff. Over the next century, similar institutions multiplied locally, as well as nationally, and would treat an increasing share of the region's ill and injured, maintaining, in the eyes of founders, those very individuals whose efforts ensured the health of the local economy.

Though Birmingham's voluntary hospitals have been investigated by historians in the past, these medical charities, like those of other communities and especially London, have generally been the subjects of single-institution studies.[14] As such, their histories are rarely presented in a way that describes their collective appearance locally or attempts to draw comparisons between institutions. While past research has detailed individual hospitals' dates of foundation, the progressive increase in the number of beds

[10] BCLA, Manuscript history of medicine in Birmingham by James Russell, *c.* 1850, uncatalogued.

[11] Tomlinson, *Medical Miscellany*, ii.

[12] E. Hopkins, *Working-Class Self-Help in Nineteenth-Century England: Responses to Industrialization* London, 1995), 10.

[13] Lane, *Social History of Medicine*, 69.

[14] See for example, G. Griffith, *History of the Free-Schools, Colleges, Hospitals, and Asylums of Birmingham* (London, 1861), 228–306; J. T. Bunce, *Birmingham General Hospital, 1779–1897* (Birmingham, 1897); R. Waterhouse, *Children in Hospital: A Hundred Years of Child Care in Birmingham* (London, 1962); J. M. Malins, 'A history of the Birmingham General Hospital', *Midland Medical Review* 11 (1975), 18–21; J. M. Malins, 'The General Hospital, Birmingham', in *Historical Sketches in the West Midlands*, vol. 2, ed. R. N. Allan and M. G. Fitzgerald (Shipston-on-Stour, 1984), 82–6; M. W. White, *Years of Caring: The Royal Orthopaedic Hospital* (Studley, 1997).

and the names of medical staff at various stages in their histories, existing studies give little indication of the ways in which these facts and figures compare and contrast with other institutions locally, let alone nationally. Neither do they emphasise periods of calamity and catastrophe from which the most important lessons are so often drawn. This book is an effort to do just this. In particular, this study seeks to summarise research that concentrates on the history of the Birmingham teaching hospitals. These include the General Hospital (established in 1779), the Orthopaedic Hospital (1817), the Eye Hospital (1823), the Queen's Hospital (1841), the Ear and Throat Infirmary (1844), the Dental Hospital (1858), the Children's Hospital (1861), the Women's Hospital (1871) and the Skin Hospital (1881). Though very different in terms of the services each originally offered, these institutions share the trait of being teaching hospitals, having been affiliated with the town's medical school – soon after its foundation in 1828 – during the late nineteenth and twentieth centuries.

Although these were not the only hospitals associated with medical education in Birmingham throughout this period, they were the earliest and, some might therefore say, most important. Some notable exceptions, however, will be apparent. For example, one important absence from this study is any description of hospitals for the mentally ill, many of whom were also housed in workhouse infirmaries, received little special treatment and were only rarely of interest to local medical educators before 1900. While donors had originally suggested building an asylum for lunatics alongside the town's general hospital in 1779, this proposal was not acted upon as was a similar proposition in nearby Leicester.[15] An asylum catering specifically to these neglected individuals eventually appeared in Birmingham in 1850; however, connections to the medical school remained limited. Its organisation was very different from the voluntary hospitals considered in this volume, and its history has been left to another study. The same applies to the town's workhouse infirmary and fever hospital, which, as many contemporaries rightly argued, would have made ideal teaching hospitals, but were not considered prestigious enough by local medical practitioners to warrant such official titles. Given these omissions, this volume can be considered an attempt to sketch out, in the broadest of strokes, a history of what have been described as an elite group among medical institutions from the belated appearance of the General Hospital in 1779 to the outbreak of the Second World War, when this and half a dozen other hospitals essentially became the core, some might even say the jewels, of a very heavily capitalised local medical service. As such, it aims to offer a short, but

[15] E. R. Frizelle and J. D. Martin, *The Leicester Royal Infirmary, 1771–1971* (Leicester, 1971), 78–95.

comprehensive, introduction to a key branch of health care in the Second City over one and a half centuries of significant social, economic and scientific change.

Finally, the volume's preoccupation with teaching hospitals also serves to broaden the study somewhat, as the educational functions of these charities extended beyond simply instructing medical students. For example, these institutions concentrated medicine and medical knowledge in a very novel manner and, in this way alone, changed the way in which concepts of health and illness were understood by local practitioners, nurses and mothers, not to mention all those fortunate and unfortunate individuals who passed through its waiting rooms and wards between 1779 and 1939.

The study is divided into two parts. Part I focuses on the emergence and development of the medical school and teaching hospitals during the nineteenth century. Chapter 1 concentrates solely on the General Hospital and begins in an earlier century, with the plans to build a hospital in 1765. It outlines the administration of the hospital in considerable detail in order to demonstrate the way in which voluntary hospitals were initially established and run throughout the nineteenth century. As a result, besides providing an overview of the early history of Birmingham's oldest and largest teaching hospital, it should serve as a model depicting the way in which Birmingham's voluntary hospitals were organised in an age preceding a government health service.

The remainder of the chapters in Part I consider two or three general and specialist institutions jointly. Chapter 2 discusses the emergence of the town's two first specialist institutions, the Orthopaedic and Eye hospitals, from very cost-effective and popular dispensaries into charities that offered the sick poor inpatient facilities, as well as medicines and minor surgical procedures. Chapter 3 considers the foundation of the medical school and the first provincial hospital built specifically for teaching, the Queen's Hospital, through their first three turbulent decades. Chapter 4 considers three very different specialist institutions, the Ear and Throat, Children's and Women's hospitals. Their varying rates of growth and diverse forms of organisation illustrate both the challenges facing medical practitioners in these years, as well as opportunities for advancement in the Victorian medical community. Chapter 5 describes the foundation of two more controversial specialist institutions, the Dental and Skin hospitals, which, over a similar period, attained more respectable, professional status. The association of a single family with the development of both specialties only further justifies their being compared in a separate section. Chapter 6 considers the state of the voluntary hospitals and their association with the medical school collectively in the last decade of the nineteenth century, when Birmingham was first recognised as England's second city. The period witnessed not only the reconstruction of the General Hospital into the premier hospital in the Midlands, but also the medical school's amalgamation with Mason's College, a liberal arts college, and the transformation of this institution into the University of Birmingham.

Part II considers the history of the voluntary hospitals and medical education in this new collective context in the first half of the twentieth century. Chapter 7 discusses the hospitals jointly up to the conclusion of the First World War. It considers the rush to rebuild following the reconstruction of the General Hospital in the last years of the nineteenth century, years which introduced many hospital governors to the hardships of hospital finance. Chapter 8 considers the state of medical education at the newly founded University of Birmingham and emphasises efforts to modernise instruction at an outdated and under-sized medical school in order to keep up with developments in practice, and a growing student population, not to mention other provincial medical schools. Chapter 9 returns to the hospitals and assesses their ability to do the same while confronting far more familiar issues, such as restricted, or in some cases declining, incomes. The length of chapters 7 and 9 reflects both the abundance of twentieth-century hospital records with an increase in committees and subcommittees, and the flurry of activity at local institutions in these decades. Chapter 10 considers the integration of medical education and clinical work in the city following the construction of a new Hospital Centre and medical school in Edgbaston. Among other things, it demonstrates the way in which instruction, while influenced by the medical licensing bodies in London remained equally determined by local conditions and concerns. A conclusion considers the main changes marking the histories of Birmingham's first nine voluntary hospitals and its medical school over nearly 200 years and draws together some of the volume's main themes.

PART I

The Emergence of
the Voluntary Hospitals

1779–1900

CHAPTER I

Birmingham's First Voluntary Hospital

NATIONALLY, Birmingham was relatively slow to construct its first voluntary hospital. Of England's provincial centres, Winchester and Bristol were the first towns to found such institutions, both of these communities having been served by hospitals from 1737.[1] By the end of the eighteenth century, another two dozen towns, of which Birmingham was one of the last, established similar medical charities.[2] Although situated in an industrial region with its numerous associated hazards, Birmingham established a general hospital only in 1779, approximately thirty years after similar institutions had appeared in Liverpool, Manchester and Newcastle.[3] Largely following a pattern that had almost become routine, the Birmingham project still managed to stand out as unusual, given that it was originally championed in 1765 by Dr John Ash, who was both an outsider and a doctor, most other institutions having been founded by local clergymen and funded by wealthy merchants, as well as ordinary shopkeepers and tradesmen.[4] As was common to many of such early charitable initiatives, meetings concerning the establishment of a 100-bed hospital in Birmingham were held at a public house, in this case the Swan Inn. William Small, a local physician, was the only other local medical man to join Ash on the organising committee.[5] Thereafter, as had occurred in nearly every community where voluntary hospitals appeared, a group of trusted and energetic supporters commenced to canvas the homes and businesses of the region's inhabitants in order to fund the venture.[6]

Not surprisingly, given the time it initially took for such a project to emerge in Birmingham, matters continued to proceed at a very slow rate, despite Ash having acquired a suitable plot of land and assembled

---

[1] R. Porter, 'The gift relation: philanthropy and provincial hospitals in eighteenth-century England', in *The Hospital in History*, ed. L. Granshaw and R. Porter (London, 1989), 150.

[2] A. Wilson, 'Conflict, consensus and charity: politics and the provincial voluntary hospitals in the eighteenth century', *English Historical Review* 111, 442 (1996), 601.

[3] Ibid., 602.

[4] Porter, 'The gift relation', 161.

[5] R. Waterhouse, 'Portrait of a medical man: Dr John Ash and his career', in *Medicine and Society in the Midlands, 1750–1950*, ed. J. Reinarz (Birmingham, 2007), 16.

[6] BCLA, GH/1/2/4.

a thirty-one-member building committee.[7] While this rate of progress was not unusually slow, there were several reasons for the dilatory manner in which matters subsequently progressed. To begin with, it was not the first medical institution in the town, a workhouse having existed in Birmingham since 1733.[8] Several of its rooms, as at 500 other workhouses in existence throughout the nation in 1750, would have been occupied by ill and infirm paupers. In response to criticisms that the parish's suffering poor were already sufficiently served by the workhouse, Ash claimed that half the town's sick came from outside Birmingham and, because they were not legally settled in the parish, were not entitled to local Poor Law services. Many other sick and injured inhabitants were denied access to such pauper institutions given their earnings and were expected to pay for the services of a medical practitioner. Nevertheless, the charity's supporters were confronted with additional evidence that clearly suggested that the presence of another medical institution was not critical, at least at that precise moment. For example, according to the records and writings of local practitioners, Birmingham was a relatively healthy environment and, as such, inhabitants may not have perceived a need for another costly charity.[9] Furthermore, the fact that other ventures were promoted as vigorously as a medical institution, including bridges, roads and even canals, only hindered the hospital committee's fund-raising efforts, especially after war broke out against the American colonies when concerns were naturally directed to a projected decline in foreign trade.[10] As a result, the charity was shelved for approximately a decade, the hospital abandoned as a boarded-up, empty shell, its grounds in subsequent years hosting a number of informal, local sporting events, including football matches.[11]

Only in 1778, with the cessation of hostilities and the completion of competing projects, were building activities resumed. Following a sustained effort in that year, the hospital was finally completed by local tradesmen, although the building had been somewhat reduced in size, opening its forty beds to patients, including the Lichfield labourer William Jones,

[7] J. Reinarz, *The Birth of a Provincial Hospital: The Early Years of the General Hospital, Birmingham, 1765–1790*, Dugdale Society Occasional Paper 55 (Stratford, 2003).

[8] *Aris's Gazette*, 18 November 1765; J. Money, *Experience and Identity: Birmingham and the West Midlands, 1760–1800* (Manchester, 1977), 11.

[9] Tomlinson, *Medical Miscellany*, 205.

[10] E. Hopkins, *Birmingham: The First Manufacturing Town in the World, 1760–1840* (London, 1989), 63; Money, *Experience and Identity*, 88, 202; Waterhouse, 'Portrait of a medical man', 16–18.

[11] For a more lengthy coverage of the hospital's early crisis, see Reinarz, *Birth of a Provincial Hospital*, and Wilson, 'Conflict, consensus and charity'.

on 2 October 1779. In the process, the once neglected institution, much like others elsewhere, such as the soon-to-be-built Derby Infirmary (1810), had become a showpiece for local trades and industries, furnished as it was with water-closets, an 'elaboratory' and iron beds, which were to be found only in a limited number of eighteenth-century provincial hospitals at this time.[12]

Like many other hospitals designed and built in these decades, the General was very domestic in appearance, looking much like a large, three-story mansion with an extensive basement. On entering the building from the front entrance on Summer Lane, 'a dirty, narrow lane' in the words of Birmingham's first historian,[13] patients and all other visitors found themselves in the main hall, facing an apothecary's dispensary and a less-than-conspicuous adjoining bedroom.[14] A corridor to the right conducted visitors to the hospital secretary's bedroom, a physician's room and a staircase leading to the wards above and a basement below. Behind the secretary's and physician's rooms and to the back of the house one passed through an anteroom before reaching a large committee room, where the governors and staff held their weekly management meetings, usually every Friday afternoon. Beyond this and to the very rear of the building on the right side lay a secretary's office and a privy. Most patients never saw the administrative section of the building unless they were called before the management committee for questioning or to complain about their medical treatment.

Unless simply collecting a prescription, most patients would have been directed to the left of the entrance hall, where the surgeon's rooms were situated. Across from these examination rooms was a ward with twelve beds and its own privy, as well as another staircase. Like that on the opposite side of the entrance hall, this stairway was wide enough to accommodate a sedan chair; it led to a chamber story comprising two main wards with room for twenty beds each, two smaller wards at the centre of the building each contained a dozen beds and two single-bed wards on either side of a ward that overlooked the entrance yard. As on the main floor, privies were located at the rear of the building at the end of the largest wards, strategically positioned to keep the building 'sweet'. The attic floor was identical to the floor below, the only difference being that the rear-facing room at the centre of the building, with capacity for six beds, was originally used as the

---

[12] J. Howard, *An Account of the Principal Lazarettos in Europe* (London, 1789), 131–6, 179, 198.

[13] Hutton, *History of Birmingham*, 258.

[14] The remainder of the hospital's description is based on drawings produced by the institution's architects, the Wyatts. BCLA, GH/3/5/988.

operating theatre, given that it was fitted with skylights, which made it the brightest room during the day.

The basement, by way of contrast, was a collection of small, dark rooms. To the front of the building were vaults and the apothecary's storage room, containing those items required for compounding pills, potions and most other prescriptions. These store rooms were framed by a laundry to the left and the matron's private room to the right. Behind the matron's bedroom lay the kitchen, scullery and pantry, with a servants' privy at the very rear. In the centre of the basement, below the apothecary's rooms, was a hot bath, a 'sweating' room for venereal patients, and another bedroom, presumably used by the house porter. Behind the laundry, in the rear left wing, lay another store room, a cold bath, a final privy and the dead room, in which the bodies of only half a dozen patients were laid out and examined in the hospital's first year. While mortuaries would progressively be removed from the main hospital building in subsequent generations, the only facilities which lay outside the hospital in its earliest years were the brewery, laundry and bakehouse, as well as a laboratory, all of which were housed in two detached structures on either side and to the front of the main block.

Like other eighteenth-century hospitals, the charity was supported largely by voluntary philanthropy, comprising primarily annual subscriptions, which for the next century averaged a guinea. While sources of finance did not vary as much as they would in the late nineteenth century, donors' motivations often did, and though these have been explored in countless histories of hospitals and philanthropy more generally, they are worth re-examining, as subscribers' motives were often subject to change with time and location, if not with each donor.[15] Besides John Ash's initial arguments in support of the hospital, perhaps the earliest and most enduring of incentives for contributing to medical charities was that rooted in the Christian tradition, the religious impulse behind much philanthropy having flourished particularly in these years as a result of a revival in the eighteenth-century church. Furthermore, Nonconformists, well represented among the middling sorts in unincorporated towns such as Birmingham, engaged in what has been described as practical Christianity, supporting, most noticeably, good works within their communities, often stimulated by a certain distrust of the state's powers.

Many other motives were of a more secular nature, though no less

---

[15] F. K. Prochaska, *Women and Philanthropy in Nineteenth-Century England* (Oxford, 1980); F. K. Prochaska, *Philanthropy and the Hospitals of London: The King's Fund, 1897–1990* (Oxford, 1992); R. Humphreys, *Sin, Organized Charity and the Poor Law in Victorian England* (Basingstoke, 1995); K. Waddington, *Charity and the London Hospitals, 1850–1898* (Woodbridge, 2000).

practical. A shortage of public buildings in these years, for example, augmented considerably the functions of non-religious institutions, such as hospitals and theatres. For this reason, the harshest critics of medical charities have often judged many philanthropic gestures to have been insensitive to the needs of the poor, regarding them instead as thinly disguised acts of self-interest.[16] Committee meetings and fund-raising events, for example, offered important opportunities for local leading lights to reinforce their social status, and provided many other upwardly mobile members of the community with regular opportunities to improve their standing, if not the possibility to rub shoulders with an established elite at a charity's annual meeting. The fact that merchants wishing to supply the hospital with goods were encouraged to become subscribers has also led many historians to question the benevolence of donors.

Additionally, such eleemosynary efforts were also used to raise the reputations of entire communities, provincial populations having supported certain charitable initiatives to foster urban development more generally. Lacking a charter, Birmingham remained notoriously behind other towns in terms of development, and many other local projects stand out as efforts to keep up with innovations pioneered elsewhere, motivated largely by civic pride.[17] For example, the fact that the inhabitants of Stafford and Northampton had erected their own hospitals by the 1760s would undoubtedly have inspired similar hospital-building efforts in Birmingham. While eventually depicted as a showpiece of local industry, if not essential viewing for all visitors to the region, hospitals remained an expression of secular welfare, most manufacturing communities having been especially eager to maintain the health of a local workforce, upon which the preservation of the region's industry and wealth depended. Political motives may also have played a part in concentrating philanthropy during these years of social unrest.[18] Finally, charity was not simply inspired by the potential power of the lower orders. Though described as a social balm, successfully drawing together individuals representing a multitude of social and political backgrounds, medical charities often benefited from a certain amount of conflict among local elites, whose generosity grew as the result of periodic struggles to control the institutions from which they derived their

---

[16] F. K. Prochaska, 'Philanthropy', in *The Cambridge Social History of Britain, 1750–1950*, vol. 3, ed. F. M. L. Thompson (Cambridge, 1990), 358–9.

[17] E. Hopkins, *Birmingham: The Making of the Second City, 1850–1939* (Stroud, 2001), 54.

[18] Lane, *Social History of Medicine*, 82; Wilson, 'Conflict, consensus and charity'; A. Wilson, 'The Birmingham General Hospital and its public, 1765–79', in *Medicine, Health and the Public Sphere in Britain, 1600–2000*, ed. S. Sturdy (London, 2002), 85–106.

status.[19] Consequently, in these and ensuing decades, an endless array of incentives, besides simply guilt and gratitude, continued to inspire private charity in Birmingham. Collectively, these varying motives ensured that the General Hospital and numerous smaller medical charities received a steady flow of income over many decades.

At all such institutions, in exchange for their monetary support, subscribers to the medical charities were permitted to recommend 'deserving' patients, namely those unable to pay for private medical treatment. Commonly excluded were infectious cases, such as those afflicted with smallpox, as well as the destitute and incurable, as their admission was seen to convert the charity into an almshouse. Rather than stress such exclusive conditions, however, the charity's printed list of nearly one hundred rules began with a far more inclusive regulation, stating that the hospital was open to the sick poor, regardless of county, the institution's management being vested in the hands of subscribers, as it was from their body that a board of governors and hospital visitors, or inspectors, were elected.[20] While all local subscribers could apply to become visitors, governors were drawn from those who subscribed more than £2 to the institution annually, each guinea donated entitling one to recommend a single inpatient, as well as an outpatient. In addition to the sums collected from subscribers, funding was directed to the charity in the form of random donations and legacies, or even fines and, famously in the case of Birmingham, musical entertainments. The first concert in support of the General Hospital was organised in 1768 at St Philip's. The event failed to reinvigorate the hospital's finances, but led to the holding of another in 1784; the festival became a triennial event thereafter.[21] Besides raising £1,325 and lasting for three days, the latter musical entertainment was staged for the benefit of the hospital exclusively.[22]

Just as the numbers of people attending music festivals continually rose in subsequent years, the number of patients admitted to the hospital increased after 1779. In the first three months of the hospital's existence, 200 patients were treated at the institution by a staff that included a matron, two nurses, four physicians, four surgeons and an apothecary. Although incomplete, the first three months of the earliest admissions register are of particular interest as they are the only eighteenth-century records for Birmingham that

[19] S. Cavallo, 'Charity, power, and patronage in eighteenth-century Italian hospitals: the case of Turin', in *The Hospital in History*, ed. L. Granshaw and R. Porter (London, 1989), 107.

[20] BCLA, GH/1/4/510.

[21] Waterhouse, 'Portrait of a medical man', 16.

[22] *Lancet*, 12 October 1823.

actually list patients' ages. Of the 127 patients admitted between 2 October and 31 December 1779, the details of 104 cases were recorded in full by the apothecary.[23] According to the register, when the patient population is divided into five-year cohorts above the age of ten, the largest group, as at other early provincial hospitals,[24] was that of men between the ages of twenty-six and thirty, comprising ten individuals, or 10 per cent of hospital admissions. The next greatest is women between the ages of twenty-one and twenty-five, accounting for nine cases. Interestingly, the hospital also admitted six children under the age of ten, despite the existence of rules precluding their admission; all such patients, as was common elsewhere, resided in the women's wards in these years.[25] The hospital appears to have been more successful at keeping out the old and infirm, recording only four patients over the age of sixty and none over seventy. In general, such patients were regarded as chronic cases and would therefore have been sent to the town's workhouse infirmary or received care from the Poor Law authorities. Other cases barred from admission included those suffering from smallpox, advanced cases of consumption, incurable cancers, or any other persons judged to be incurable, a policy that generally permitted voluntary hospitals to increase patient turnover and present very favourable reports to subscribers.[26] Other persons regularly excluded from voluntary hospitals included pregnant women and most children under the age of seven, two groups generally considered to be highly infectious, as well as those individuals found to be 'disordered in their senses'.[27] Quite predictably, medical charities aimed specifically at these neglected members of the local population would be established in subsequent decades.

An effort to categorise patients by illness, while interesting, actually reveals very little reliable medical data, given the state of disease nomenclature and variations between individual practitioners' methods of diagnosis during these years. Nevertheless, if one were to sort cases according to available information, the greatest number of men treated as inpatients appear to have suffered from sore or ulcerous legs, as was indicated in eight separate cases, only one more than accidents, the next greatest

[23] BCLA, GH/4/2/560.

[24] A. Berry, 'Patronage, Funding and the Hospital Patient, *c.* 1750–1815: Three English Regional Case Studies' (DPhil diss., Oxford University, 1995), 221–5.

[25] B. Abel-Smith, *The Hospitals, 1800–1948* (Cambridge, MA, 1964), 24–5; J. A. Walker Smith, 'Children in hospital', in *Western Medicine: An Illustrated History*, ed. I. Loudon (Oxford, 1997), 222.

[26] BCLA, GH/1/4/510.

[27] BCLA, GH/21/4/510; Lane, *Social History of Medicine*, 85–6; Porter, 'The gift relation', 165; A. Rook, M. Carleton and W. G. Cannon, *The History of Addenbrooke's Hospital* (Cambridge, 1991), 46.

category.[28] Furthermore, approximately five patients each were designated as either scrofulous, consumptive, lame, rheumatic, venereal, or suffering with ulcer. Two interesting illnesses noted among the remaining isolated cases include diagnoses of leprosy and hypochondria. All other remaining cases represent various symptoms rather than actual diseases, including cough, sore eyes, pain in the bowel and bones and dyspepsia. With the exception of two amputations of the knee, both patients having survived their operations, treatment appears to have been conservative.[29] A notable exception to the rule was Elizabeth Suffolk, who was discharged by medical staff in March 1788 after refusing to have her hand amputated, one of many reminders that patients were far from being the passive recipients of health care many have imagined.[30]

While accidents were very rare among women, the overwhelming complaint of female patients was 'sore legs', with fifteen cases recorded in the first patient register. Though this may seem surprising, since such cases have rarely been mentioned by medical historians, leg ulcers were regularly addressed in the writings of eighteenth-century medical practitioners, who estimated that one in five labourers suffered from this affliction.[31] At other provincial hospitals, these cases comprised between 5 and 20 per cent of patients, and a charitable institution devoted solely to the treatment of leg ulcers was even mooted in these years.[32] At Birmingham, where the figure reached 22 per cent in the hospital's first decades, leg ulcers overshadowed all other ailments. Though less common in nineteenth-century hospital

---

[28] BCLA, GH/4/2/560.

[29] BCLA, GH/1/4/510; Tomlinson, *Medical Miscellany*, 71. Treatment in Manchester appears to have been similarly conservative; see J. Pickstone, *Medicine and Industrial Society: A History of Hospital Development in Manchester and its Region* (Manchester, 1985), 12; see also S. Cherry, 'The hospitals and population growth, Part 1: The voluntary general hospitals, mortality and local populations in the English provinces in the eighteenth and nineteenth centuries', *Population Studies* (1980), 258.

[30] BCLA, GH/1/2/5; A. Tomkins, *The Experience of Urban Poverty, 1723–82: Parish, Charity and Credit* (Manchester, 2006), 121.

[31] I. S. L. Loudon, 'Leg ulcers in the eighteenth and early nineteenth centuries', *Journal of the Royal College of General Practitioners* 31 (1981), 263; some eighteenth-century sources include: B. Bell, *A Treatise on the Theory and Management of Ulcers* (Edinburgh, 1777); W. Rowley, *A Treatise on the Cure of Ulcerated Legs without Rest* (London, 1774); and M. Underwood, *Surgical Tracts, including a Treatise upon Ulcers of the Legs* (London, 1788); [The Inquirer], 'On the treatment of ulcerated legs', *Edinburgh Medical and Surgical Journal* 1 (April 1805),187–93.

[32] Loudon, 'Leg ulcers', 263–4; G. Risse, *Hospital Life in Enlightenment Scotland* (Cambridge, 1986), 138–40.

registers, such cases, if one were to rely on the memoirs of provincial practice, continued to comprise a sizeable proportion of total cases attended by private practitioners. Fever, consumptive and hysterical cases, on the other hand, each appear only three times among women treated at the General Hospital, while all other categories of illness, including asthma, worms, piles, rheumatism, constipation and venereal disease, are recorded on no more than one or two occasions. As valuable as the earliest patient data are, diseases were subject to significant change from year to year. In 1780, for example, of the 846 cases treated, the patient population included only four hysterical cases, six patients with worms and three with piles. Fever, asthma and consumptive cases, on the other hand, climbed significantly, numbering twenty-two, thirty-five and forty-eight respectively, while sore legs continued to account for a tenth of diagnoses.

In general, daily treatment at the institution would have become quite routine, comprising primarily the prescription of many medicines of vegetable and mineral origin in the form of powders, boluses and lozenges, most of which would also have been prepared on the hospital premises. With the help of an assistant, in many cases an apprentice, the hospital's apothecary prepared the appropriate draughts, juleps, mixtures and elixirs, embrocations, liniments, salves and ointments, as well as a wide variety of drops, cordials, electuaries and linctuses, depending on patients' salient symptoms. For example, should the need for expectorants have been determined, garlic, tobacco or quicksilver might have been prescribed, or in the case of diuretics, onion, elm, or, most famously, foxglove.

In general, early medical preparations can be divided into sedatives and stimulants, as well as a number of common physical treatments. Each week medical staff would apply and reapply plaster poultices, fomentations and dressings, while as frequently carrying out the most common depletive therapies of the day, such as bleeding, by way of lancet or leech, blistering and purging. Equally common in the late eighteenth and early nineteenth centuries was vaccination, which medical staff carried out at the institution each Tuesday at 3 pm free of charge, nearly a dozen parents having brought their children to the hospital for this purpose each week. On average, 450 children were vaccinated each year between 1801 and 1815.[33] During the following decade, numbers almost doubled, though continued to be outnumbered by those cases that came to be bled and purged. Less common surgical procedures included the manipulation of dislocated bones, the tapping of hydroceles, treatment of fistulas and aneurism, drainage of abscesses, reduction and setting of fractures, or the performance of amputations, eleven cases of the latter procedure being undertaken in the hospital's

[33] BCLA, General Hospital, Annual Reports, 1801–15, MS 1921/414.

second year.[34] Throughout these first years, each medical officer treated approximately 170 patients annually, or fourteen per month, a case load that was easily combined with the demands of private practice. Nevertheless, physicians and surgeons at all voluntary hospitals occasionally needed to be reminded to attend patients regularly in these years.

Along with the Oxford-educated Ash, the hospital's first physicians included the well-known William Withering, one of the most eminent English botanists of his time, who had previously for eight years been physician to the infirmary in Stafford.[35] These two eminent medical men were joined by the relatively unknown Thomas Smith and Edward Johnstone. The last named, though having only just received his doctorate, stemmed from a family that had provided Worcestershire and Warwickshire with physicians for more than a century. The hospital's four surgeons included one of the few natives of the town, Robert Ward, who, through 'the useful qualification of small talk' and extensive engagement in midwifery, cultivated one of the largest private practices in the region.[36] He was joined by George Kennedy and John Freer, both of nearby Atherstone, the latter having trained for a time in London with John Hunter. The fourth and final surgeon, Jeremiah Vaux, also trained in London before inoculating according to the Suttonian method many of the inhabitants of Herefordshire and Monmouthshire. A member of the Society of Friends, Vaux subsequently settled in Birmingham, where several wealthy and influential Quaker families formed the nucleus of his practice.

Interestingly, few of these local practitioners, though known to us today, are ever mentioned in hospital minute books besides the odd reference to their attendance at meetings, most having made only one or two appearances at the institution weekly. In turn, the published works of these practitioners mention the hospital only rarely. Withering's treatise on the foxglove is an exception; it mentions half a dozen patients who were relieved by his botanical treatment.[37] Attending the hospital as frequently as the physicians was Richard Sims, a barber, who was paid 10s 6d quarterly to shave the patients twice a week. Besides attending the General Hospital every Thursday to visit patients, and one Saturday in the month to discharge any inmates judged to be healthy, medical staff were also recognised as house governors when in the charity's service. Unlike their counterparts

---

[34] E. Alanson, *Practical Observations on Amputation* (London, 1782).

[35] J. K. Aronson, 'William Withering', in *Oxford Dictionary of National Biography*.

[36] BCLA, Manuscript history of medicine in Birmingham by James Russell, *c.* 1850, uncatalogued.

[37] W. Withering, *An Account of the Foxglove and Some of its Medical Uses: with Practical Remarks on Dropsy and Other Diseases* (London, 1785).

at many other institutions, all of the first medical officers also donated substantial sums of money to the hospital, except for Vaux, who nevertheless used his connections to attract additional funding for the institution during the twenty-eight years he held the post of surgeon.[38] In general, medical officers held their coveted posts for considerable periods; only twelve physicians and ten surgeons served the charity during its first three decades. On average, physicians held their posts for sixteen years, six years less than the hospital's more committed surgeons. At other provincial hospitals, surgeons similarly served on average a third longer than physicians, who seemed more interested in building lucrative private practices than serving charitable institutions for decades.[39] That said, physicians would have found it easier than surgeons to build large private practices in the first place.

Though the General Hospital in Birmingham was described as a hospital for the Midlands counties, it is not clear from early minute books and publicity material if this was actually the case in these years. If one were to rely on patient registers for 1779 and 1780, for example, this implied primarily Warwickshire, Staffordshire and Worcestershire. Nevertheless, of more than 600 patients for whom staff recorded a parish of origin in these years, and despite John Ash's earliest printed defence of the institution, two-thirds were recorded as residents of Birmingham, and another eighty-five, or 14 per cent of inmates, originated from one of twenty-four other towns and villages in Warwickshire, most giving Aston and Sutton Coldfield as their parishes of origin. Staffordshire accounted for seventy-one, or nearly 12 per cent of patients, most coming from Walsall, Handsworth and Darlaston. Only a single patient came from Stafford itself, a fact that should not be surprising, given that the town opened its own general hospital in 1766. Less than 4 per cent, or twenty-five of the General's patients, originated from Worcestershire, again the majority residing in nearby parishes, such as Halesowen, rather than Worcester or Stourbridge. The remaining twenty-seven, or 4.4 per cent of patients, came from approximately a dozen other

---

[38] Porter, 'The gift relation', 161; Hilary Marland, *Medicine and Society in Wakefield and Huddersfield* (Cambridge, 1987), 118.

[39] See, for example, J. D. Leader, *Sheffield General Infirmary* (Sheffield, 1897), 155–6; G. Munro Smith, *A History of Bristol Royal Infirmary* (London, 1917), 483–5; F. H. Jacob, *A History of the General Hospital near Nottingham* (Bristol, 1951), 335–6; G. H. Hume, *The History of the Newcastle Infirmary* (Newcastle-upon-Tyne, 1906), 142–3; W. H. McMenemey, *A History of the Worcester Royal Infirmary* (London, 1947), 326–7; F. F. Waddy, *A History of Northampton General Hospital* (Northampton, 1974), 154–60. On average, the 90 physicians and 53 surgeons who served at the general hospitals in Bristol, Newcastle, Northampton and Worcester until 1810 and Nottingham and Sheffield until 1820 retained their offices on average for 13 and 24 years respectively.

parishes from London to Inverness.[40] In eighteenth-century Birmingham, as in other towns and cities, hospital governors appear to have 'talked up' their catchment areas for publicity purposes.[41]

The cost of providing care to nearly 600 patients in the hospital's first year approached £1,500 and remained at this level for more than a decade. Of this annual expenditure, the largest single sum, approximately £230, or 16 per cent, was spent on food, the hospital's diet largely comprising, depending on the time of day and year, bread and butter, meat and potatoes and cheese and beer. Another £160 was spent on drugs and the raw ingredients dispensed by the apothecary in various forms; only £66 (5 per cent) was paid in salaries, half of this going to the apothecary, the other half largely representing cleaning and nursing expenses until the middle of the nineteenth century. Of general household costs, the majority was spent on coals (£35), or heating the hospital's wards, while £16 went towards buying the soap and candles that cleaned and illuminated the institution. With expenses rarely exceeding £2,000, the hospital appeared to offer the charity's supporters value for money. By way of comparison, in the late-1780s, the town's workhouse and its 300 inmates on average cost the local rate-payers £14,000 annually, of which £450 (or 3.2 per cent) went towards medical expenses. In most years, between forty and seventy-five paupers were to be found in the infirmary, women almost always outnumbering men. In addition, the majority of the workhouse infirmary's medical costs went towards the care and treatment of lunatics (£234), as well as the purchase of drugs (£133).

With an increase in donations over the next decade, the General Hospital's forty beds gradually increased to one hundred. Again, donations, such as a particularly generous one from the merchant Samuel Garbett, filled empty wards with fully dressed beds, and an extension was eventually built in 1792. Despite this impressive rate of growth, expenses continued regularly to exceed income, but the charity still only cost its supporters a tenth of the amount spent on the local workhouse. As interestingly, three years later, in 1795, the hospital was annually treating nearly 1,400 patients, who often appeared to differ little from those treated at the workhouse infirmary. While it has been suggested that the hospital was built for patients working though without legal settlement in the parish, 82 per cent of cases came from forty-two Warwickshire parishes, 913 alone (or 68 per cent of all patients) from the town of Birmingham. Cases

---

[40] BCLA, GH/4/2/560.

[41] M. Fissell, *Patients, Power and the Poor in Eighteenth-Century Bristol* (Cambridge, 1991), 96; G. Mooney, B. Luckin and A. Tanner, 'Patient pathways: solving the problem of institutional mortality in London during the later nineteenth century', *Social History of Medicine* 12, 2 (1999), 227–69.

from Aston more than doubled since the hospital opened, reaching sixty-six, significant increases in patients from Solihull are equally noticeable. Though cases from Bilston, Handsworth and Walsall had also increased appreciably, thirty-three cases, more than a fifth of Staffordshire patients, originated from West Bromwich, most having been recommended to the institution by Lord and Lady Dartmouth; another twenty came from Harborne. As a result, the number of patients from the county remained steady, comprising 11.5 per cent of all individuals treated at the institution. Cases from Worcestershire, on the other hand, had declined noticeably since 1779, most coming from nearby Yardley; only 2.6 per cent of patients came from this neighbouring county, whose inhabitants had been supporting a hospital for more than three decades. Those from all other regions in England remained static, comprising 2 per cent of cases, due only to the large numbers of soldiers treated at the General during these revolutionary years. The first set of hospital regulations indicates that soldiers were indeed anticipated at the institution from the charity's inception. The fact that they were charged half a crown (2s 6d) a week for subsistence also makes them the first paying patients treated at English voluntary hospitals.

During these years, staff treated between 600 and 1,000 inpatients and an equal number of outpatients. From 1812, however, outpatients proceeded to increase more rapidly, exceeding inpatients numbers, which remained static, by a multiple of three by 1820. Although patient registers for this period survive and depict a varied array of illnesses and injuries, if one relied solely upon annual reports and minute books to reconstruct the work of medical staff in these years, one might easily become convinced that the hospital treated mainly drowned victims. At times, minutes of meetings recount little else than the number of children who were fished from the town's newly built canals along with the rewards paid to their rescuers. While the building of the first canal in Birmingham clearly delayed the construction of the hospital, the presence of these artificial rivers in the region hastened the appearance of drowned victims, and especially children, in the hospital's registers, largely because so many were falling into these waterways. So long used to a canal-free Birmingham, many of the town's inhabitants took some time to become accustomed to the town's new water hazards. Others turned to the canals as their chosen method of suicide.

As a result, at the hospital's anniversary dinner in 1803, governors decided to institute the plans of the Humane Society, first outlined in London in 1790, in order to aid the recovery of persons apparently dead from drowning and other accidents, alongside those of their charity. As such, the hospital became responsible for administering the work of the Humane Society locally, while subscribers were encouraged to display instructions for reviving victims of drowning in their homes, and otherwise distribute similar leaflets among their neighbours, public houses and all navigation offices. Proper houses were appointed near the town's canals, where drugs and the necessary reviving apparatus were deposited for the use of the town's lay paramedics. A reward was given to those occupying a house other than those specifically set aside, as well as to those who assisted the rescue of individuals. Persons who simply notified the occupants of a designated house or a medical man of a suspected drowning received 2s 6d, while those who helped fish a body from the waterways and attempted to revive victims for a period of four hours received a guinea, usually divided equally among the rescuers. Each housekeeper who offered the necessary accommodation for such recoveries was also offered a guinea, but was excused responsibility for burial expenses in an unsuccessful case. Their co-operation was almost certainly guaranteed as an additional guinea was offered in all successful cases. As at the hospital, the assistance of medical men was honorary, though any who helped revive persons thought dead received the Humane Society's silver medal. As a result, in the first year, the hospital spent an additional £3 16s printing notices describing the society's instructions to restore persons apparently dead from drowning. That same year, a number of heroic individuals managed to rescue thirteen individuals from the town's waterways. In 1806, Mr Rabone was requested by the governors to wait upon Mr Boulton at his famous Soho works respecting a medal they wished to have cast in order 'to express their gratitude and obligation as a benevolent society to any person found useful in saving the life of a fellow creature'.[42] Over the next two decades, minute books record the rescue of at least another 140 men, women and especially children. Not surprisingly, given these numbers, this aspect of the hospital's work was advertised as frequently as any other medical services in annual reports during these years.

In contrast to reports of dramatic water rescues, the hospital's other official business appears unusually ordinary. Generally, minute books in the first decades of the nineteenth century describe a very domestic routine at weekly meetings. Bread, meat and drugs were put to tender, and the wards were whitewashed each year before the annual general meeting, when it was expected numerous governors, besides the two officially appointed visitors, would tour the building. Other matters recorded include the regular monthly admission and discharge of patients and less frequently, though equally regularly, turnovers among members of staff, especially the nurses.

[42] BCLA, GH/1/2/5.

Hospital finances regularly failed to cover all expenditures in these years. Occasionally, this led the governors to sell investments, but usually any financial shortfall also encouraged new funding initiatives. For example, in 1806, the hospital established a choral society in order to save the expense of hiring trained singers from a distance for future triennial festivals. Between festival years, the managers of the society proposed musical performances at local churches, usually St Philip's, to relieve the hospital of extraordinary expenses. Other initiatives, though less musical, relied as much on the co-operation of the clergy. For example, the following year, 1807, members of the clergy in the parishes in Warwickshire were solicited to give sermons for the benefit of the hospital. Nevertheless, until methods of co-ordinating such initiatives improved in the next half century, the music festival alone continued to grow in size and importance. In fact, so crucial to the success of the festival in these years did governors regard its chief administrator, Joseph Moore, that, in 1812, shortly after another series of three-day concerts had produced profits totalling £3,629, a special meeting was held to acknowledge his charitable work.[43]

While thanked on a tri-annual basis in the pages of the hospital's minute books, on this occasion, Moore was presented with a silver tureen along with a stand and four silver dishes, the gift having been purchased by private subscription from the Soho Manufactory. In return, the Soho managers, very interestingly, provided the hospital with workplace fines totalling £50 the following year. Before Moore took over the festival's management thirteen years previously, receipts totalled exactly £2,043. In October 1812, gross receipts reached £6,680, of which £1,600 was realised from a performance of Handel's *Messiah*, a festival favourite, on the second morning. Though much of this was due to an increase in ticket prices, which had risen from 3 to 6 shillings in 1799 to between 10 and 20 shillings, this administrative detail was overlooked on this occasion. Nevertheless, besides attaining mere financial success, Moore, an amateur musician, was said to have raised the musical meetings from 'respectable country meetings' to 'unrivalled … national grandeur and celebrity', a claim that was engraved on his tureen. Unlike previous festivals, that held in 1811 was opened with a sermon by the Bishop of Worcester and included performances by Madame Catalani, the antelope-faced prima donna who made her English opera debut in 1806, and a chorus of more than 200 performers.[44] Furthermore, Moore's efforts were to have 'infused into the lower classes, in which the physical strength of the community consists, a love of their

---

[43] BCLA, GH/1/2/5.

[44] Bunce, *Birmingham General Hospital*, 82.

country'.[45] Few would have opposed Moore's suggestion that subsequent musical meetings be 'conducted on the grandest possible scale'.[46]

Given the reputed importance of the triennial festivals, the affairs of the hospital committee continued to be preoccupied with the event's organisation. Members eventually agreed in 1820 to purchase the theatre in which the main musical performances were held. While the festival had been extended to four days, ownership of the theatre ensured future performances would not be postponed by the venue's other scheduled performances and commitments, as had frequently occurred in the past. As expected, a grander festival also produced greater profits, which now totalled £5,000. By comparison, the following year, income from subscriptions had noticeably declined in significance, comprising only 30 per cent of the institution's total income, which surpassed £5,500.

In all probability, the care provided on the hospital's wards, like its festivals, continued to improve in these years, despite relatively few changes in the sort of medicine that honorary staff practised. For example, the governing committee had by this time decided that nurses should devote themselves entirely to the care of patients, not the cleaning of wards.[47] In general, the matron was now supported by nine maids, two night nurses, three house maids, a cook and a kitchen maid. As importantly, the hospital and its staff were more carefully scrutinised by the visitors, whose duties had also been made 'more regular, systematic, and efficient'. As a result, not only were many more 'irregularities' noted at meetings, but the quality of meat and beer served at the hospital appears to have improved, and damaged items of furniture and fittings were usually repaired more quickly than in the past. Regarded by subscribers as crucial to assessing the work of the hospital, the death rate in these years averaged only 1 per cent, a remarkable figure considering that, by 1825, the hospital building had not been repaired for nearly half a century. Not surprisingly, the committee also finally began to consider the institution's extension at this time.

Alterations commenced with the introduction of gas to the building in 1826, but reconstruction was already years behind schedule. For example, two years had passed since the addition of fever wards had first been suggested to the hospital's subscribers, fever cases having been admitted to hospital only when 'proper separate accommodation could be provided for such patients'. As a result, the institution's small isolation wards were also often used by surgical cases or those suffering from 'disgusting infirmities, by the offensiveness of their sores, by their moanings under pain, or their

---

45 BCLA, GH/1/2/5.

46 Bunce, *Birmingham General Hospital*, 84.

47 BCLA, GH/1/2/7.

noise in delerium' and were usually not free when required. Given frequent outbreaks of erysipelas and other common hospital fevers, the outbreak of which could greatly reduce the work of the charity, separate wards were clearly needed for infectious cases. Aware that the charity was trailing developments pioneered elsewhere, the hospital committee looked beyond the town for inspiration, admiring in particular the kingdom's most extensive fever wards at Glasgow, as well as those at Derby, which were similarly attached to the main building, with access through a separate entrance from without. The benefits of such a design were that, should the hospital have required additional space in future, an extension could easily be provided by removing the walls of the new fever wards. Moreover, until this occurred, only two nurses were required to administer the new wards, thus limiting any associated expenses.

By 1830, following much work and the highest recorded profits of a music festival to date (£5,964), the hospital was said to be in a perfect state of repair. Moreover, since 1828, the hospital governors gained an opportunity to publicise the work of their institution more effectively than in previous years, given that the annual report, formerly a single printed sheet, had become 'a book', albeit only a few pages in length. As with many previous innovations, reports were compared with those of institutions in Manchester, Sheffield and Bristol, a practice that ensured such publicity material became very uniform at hospitals across the country, even if administrators' accounting methods varied enormously. So, too, did the work of the institution continue to vary. Outpatients surpassed 4,000, while inpatients remained nearer to 1,500 in number. After another year of increasing workloads, further improvements at the hospital appeared to signal a healthier future. In 1831, for example, the committee applied to the local waterworks, eager to diminish their reliance on well water.[48] Fortunately, this decision was reached a year before the country experienced its first outbreak of Asiatic cholera.

Even without such epidemics, the numbers of outpatients only increased in subsequent years, and hospital fevers, rather than cholera, continued to threaten the hospital with closure during this period. Nevertheless, the closure of wards, whether the result of infection or insufficient funds, would also affect the work of the charity less than it had in the past, largely because the majority of patients entering the hospital did not inevitably enter its beds, but merely attended the dispensary. During the institution's first five decades, the treatment of outpatients had proven far less a financial burden, allowing as it did hospital administrators to increase the work of the hospital at no great cost and, in this way, secure the public's long-term

[48] BCLA, GH/1/2/9.

support. More importantly, given that the existence of a general hospital, not to mention many other new municipal services, reduced the funds available to subsequent medical charities, the dispensary model would become very popular in Birmingham and elsewhere. This was especially the case after the mid-eighteenth century, when a number of dispensaries in the capital, as well as Bristol (1747), Stroud (1755) and Leicester (1776), demonstrated the unsatisfied demand for non-residential medical attention nationally.[49] In London alone, fifteen dispensaries had been founded by 1792.[50]

Founded on the same principles as an institution in Westminster, the dispensary movement in Birmingham commenced in 1792 with the foundation of a dispensary in Temple Row, in a house next to the church that would later become the town's cathedral. In its first year, the institution's first medical staff saw more than 200 cases. This figure quickly rose above 1,000 in 1801, much of the charity's finances coming from Matthew Boulton, who was elected its first treasurer.[51] By the end of 1806, the dispensary was treating more than 1,500 patients annually, comprising approximately 200 midwifery cases, and another 887 smallpox inoculations, costing only £48.[52] Not surprisingly, suggestions for a new building at Union Street were soon advanced, and a plan drawn by William Hollins was exhibited at the offices of *Aris's Birmingham Gazette*, the town's main periodical. Only days later, a foundation stone had been laid, and by February the new dispensary was treating nearly 400 of Birmingham's sick inhabitants a week.

In the nineteenth century, many more dispensaries were founded across the country. Additional dispensaries in Birmingham were opened in Highgate (1871), Nechells (1876) and Ladywood (1883). Once seen to be providing value for money, the General Hospital's costs began to concern hospital governors in subsequent years. The impact of this new, cost-effective form of medical organisation, however, was far greater than this alone suggests. 'Less splendid and expensive than hospitals', dispensaries were well adapted to towns of moderate size and afforded the widest possible sphere for the treatment of diseases, including those 'which are so infectious, malignant, and fatal, as to be excluded from admission

---

[49] Lane, *Social History of Medicine*, 89–90; Z. Cope, 'The history of the dispensary movement', in *The Evolution of the Hospitals in Britain*, ed. F. N. L. Poynter (London, 1967), 73.

[50] R. G. Hodgkinson, *The Origins of the National Health Service* (London, 1967), 209; H. Marland, *Doncaster Dispensary, 1792–1867: Sickness, Charity and Society* (Doncaster, 1989), 15.

[51] J. E. Jones, *A History of the Hospitals and other Charities of Birmingham* (Birmingham, 1908), 31.

[52] *Aris's Gazette*, 17 November 1806.

into Infirmaries'.[53] So popular did this model become in a very short time that the next generation of voluntary hospitals began as dispensaries, often operating on very limited budgets years before their administrators contemplated establishing even the most rudimentary inpatient facilities. At the same time, these institutions created many more opportunities for other local practitioners, who desired to occupy hospital posts. It is the first of these new institutions that the next chapter will address.

[53] T. Percival, *Medical Ethics* (Manchester, 1803), 23.

CHAPTER 2

Optic and Orthopaedic Charities: Birmingham's First Specialist Hospitals

WHILE THE VALUE of specialisation was recognised by many industrialists in nineteenth-century Birmingham, very few medical practitioners expounded on the beneficial effects of a division of labour in these years. Although some specialisation in medicine was occurring at this time, the majority of the medical profession initially responded to this new tendency with open hostility.[1] In general, those who relinquished *general* practice and devoted themselves to the diseases of a particular organ, such as the eye, risked being ostracised as quacks by their colleagues. Many medical practitioners consequently responded to specialisation unfavourably. Besides being considered unprofessional, most regarded it as threatening for a variety of reasons. For example, it promised to make an already fragmented medical profession, originally divided into physicians, surgeons, apothecaries, then, by the early nineteenth century, into general practitioners and consultants, even more chaotic and complex.[2] It also potentially increased the number of medical instructors who claimed specialist knowledge with whom more orthodox practitioners would potentially have to share medical students' fees.[3] Similarly, specialist institutions created throughout the Victorian period would face accusations of diverting charitable donations away from general hospitals.

In contrast, the public was anything but hostile, and patronised specialists with enthusiasm, especially since general hospitals had excluded certain diseases and age groups from their remit from their first appearance. On the other hand, medical practitioners, just like those who specialised in manufacture, began to realise that 'they could often do more good, earn more money and have less arduous work to perform' by devoting themselves to just one branch of the profession.[4] This was especially true in the provinces, where doctors did not have access to the same concentration of wealthy patrons as existed in a large metropolis, such as

---

[1] G. Rosen, 'Changing attitudes of the medical profession to specialization', *Bulletin of the History of Medicine* 12 (1942), 343.

[2] Ibid., 26.

[3] G. Weisz, *Divide and Conquer: A Comparative History of Medical Specialization* (Oxford, 2006), 27.

[4] Rosen, 'Changing attitudes', 352.

London.[5] The emergence of localised pathology in the nineteenth century additionally furnished an intellectual rationale for specialisation, though one that is not accepted by all historians.[6] Numerous other ideas and social forces, however, made specialisation appear natural in these years, not least the emergence of Darwin's theory of evolution.[7] Frequently driven by technological innovation, specialisation also increased knowledge and improved technique. Consequently, opposition to this tendency was doomed to defeat, and specialisation even attained a certain degree of respectability by the second half of the nineteenth century. In any case, the readiness with which this innovation was accepted largely depended on its close resemblance to traditional medicine. Rather than depicting themselves as alternative healers, most specialists continued to assert that specialisation was inseparable from general medicine. Its practitioners also tended to be based in hospitals, not in public houses and country markets, long associated with the business of unorthodox medical wares, including the peddling and promotion of quack nostrums.

To a number of historians, a key factor that encouraged specialisation was urbanisation.[8] Between the completion of the General Hospital and the second decade of the nineteenth century, Birmingham's population leapt from little more than 40,000 to approximately 100,000. The successful division of labour in any trade and profession depends on the increase of a local population to a size that supports such specialisation. The better management of large urban populations was also achieved through improved systems of classification. It is for this reason that specialisation in medicine has been linked with 'emerging notions of administrative rationality in the nineteenth-century state'.[9] It therefore first appeared to emerge in places such as nineteenth-century Austria, France and Germany, countries where inhabitants were long familiar with increased levels of government regulation and control. According to George Weisz, countries with strong medical associations were similarly the first to resolve the issue of specialist certification, something that would be achieved in Britain only in the twentieth century.[10] Immigration to towns, like Birmingham, however, occasionally encouraged the spread of ideas and people and equally

---

[5] S. Sturdy and R. Cooter, 'Science, scientific management, and the transformation of medicine in Britain, *c.* 1870–1950', *History of Science* 36, 114 (1998), 427.

[6] G. Weisz, *Divide and Conquer*, xix.

[7] Ibid., xx.

[8] Ibid., xiii.

[9] Ibid.

[10] Ibid., xxvi.

fostered the growth of specialisation in less strictly regulated places. The entrepreneurial and individualistic orientation of many specialists, in some respects, made them appear particularly suited to Britain. As a result, although specialist societies and certification would take some additional time to develop in England, specialist hospitals multiplied throughout the country during this period.

During the first half of the nineteenth century, more than two dozen specialist hospitals and dispensaries existed in London, one dozen having survived since the eighteenth century.[11] Another twenty-two appeared in various provincial towns and cities. Institutions whose founders devoted themselves to the diseases of the eye were noticeably prominent during these years. In both Britain and on the Continent, ophthalmology is generally regarded to have constituted the first modern specialty.[12] In some respects, therefore, Birmingham appears typical when compared with national and even international trends in specialisation. One of the first institutional representatives of this new medical trend in Birmingham was an eye hospital. The very first, however, was an orthopaedic medical charity and, therefore, a more unusual start to specialisation in local medical practice.

SPECIALISATION IN ITS INFANCY: THE ORTHOPAEDIC HOSPITAL

For the remainder of the eighteenth century and the first decade of the nineteenth, the General Hospital and the town's General Dispensary were the sole voluntary medical charities in Birmingham. By 1817, however, a very different and specialised institution appeared. Founded by George Freer, a local surgeon with a national reputation due his work on aneurism (1807), the Orthopaedic Hospital opened on 24 June 1817, its medical officers initially dispensing advice and apparatus from a building on New Street. Inaugurated as 'The Institution for the Relief of Hernia, Club Feet, Spinal Diseases, and all Bodily Deformities', the hospital was very much like other specialist institutions formed in these years in its governors' decision to name the hospital in a manner that clearly described the type of patients staff intended to treat. Hospitals were, after all, relatively new institutions, and this decision was intended to avert much confusion about the charity's specific function. For example, St Mark's Hospital in London was founded almost two decades later as the 'Benevolent Dispensary for the Relief of the Poor Afflicted with Fistula, Piles and other Diseases of the Rectum and Lower

---

[11] Ibid., 29.

[12] Ibid., 30.

Intestines'.[13] In the case of the Orthopaedic Hospital, this strategy clearly worked. In the following decades, as the number of registered patients and subscribers expanded, only the hospital's name contracted. Expansion, however, proceeded in a very gradual and cautious manner, staff having increased premises only marginally with each move, first into the home of one of its surgeons in 1824, then its own building on Great Charles Street in 1857.

In many ways, the decision to establish an orthopaedic charity was an unusual choice for the town's second voluntary hospital; in most other regions, early specialist institutions were either lying-in (maternity) or lock (venereal) hospitals.[14] By choosing to promote an alternative institution, Birmingham's medical community created one of the earliest orthopaedic hospitals in Europe, being pre-dated by a similar institution founded by Jean Venel thirty-seven years earlier (1780) in Orbe, Switzerland.[15] Nicholas Andry had coined the term that defined the speciality only four decades earlier when the Parisian physician's textbook, *Orthopaedia* (1741), first appeared. Given that Birmingham was located in an industrial region, the choice of an orthopaedic charity might have appeared exceptionally well suited to the town and its local economy in the early nineteenth century. Industrialisation, it has been argued, excluded impaired people who had previously been integrated and socially active members of their communities, and the earliest orthopaedic surgeons sought to reintegrate these individuals into productive society.[16] However, the first cases treated at the Birmingham hospital were not adult workers disabled by their industrial employments, but local children born with malformed limbs. In this respect, perhaps more influential in the charity's establishment was the work undertaken by Robert Chessher practising some 40 miles outside Birmingham in Hinckley, Leicestershire. Almost the only instance of a purely orthopaedic surgeon in the eighteenth century, Chessher devoted himself to the management of spine and limb deformities soon after resigning as house surgeon at the Middlesex Hospital and returning to Hinckley in 1778.[17] By 1810, he was treating the bones and joints of 200 young patients with friction and motion,[18] not to mention splints

---

[13] See L. Granshaw, *St Mark's Hospital, London: A Social History of a Specialist Hospital* (London, 1985), 1.

[14] R. Porter, *The Greatest Benefit to Mankind* (London, 1997), 298.

[15] R. T. Austin, *Robert Chessher of Hinckley, 1750–1831: The First English Orthopaedist* (Leicester, 1981), 7.

[16] A. Borsay, *Disability and Social Policy in Britain since 1750* (Basingstoke, 2005), 11.

[17] D. LeVay, *The History of Orthopaedics* (Carnforth, 1990), 87.

[18] B. Valentin, 'Robert Chessher (1750–1831): an English pioneer in orthopaedics', *Medical History* 2 (1958), 309.

and appliances, many of which he invented, including the double-inclined plane which so impressed Sir Astley Cooper.[19] Over time, Chessher's methods gained him 'a greater degree of reputation than has attached to any other individual in the same patients. Many [p]ersons in the highest ranks of life did not hesitate to commit their children to the professional care of the eminent surgeon'.[20] Though misspelling his name, George Eliot reinforced his legacy in *Middlemarch*, referring to 'Mr Cheshire, with his irons, trying to make people straight when the Almighty had made them crooked.'[21] Offering such radical cures, his fame could not have bypassed Birmingham medical society, especially since John Freer, brother to the founder of the Birmingham orthopaedic charity, was, like Chessher, a pupil of John Hunter. If not personally acquainted with the Hinckley surgeon, George Freer would surely have noted Chessher's success, not to mention other important works on orthopaedics published by contemporaries, including Percival Pott's investigations into spinal deformity.[22] Importantly, Freer was particularly eager to improve his credentials following a controversial appointment to the General Hospital in 1793 at the age of twenty-one, following the death of his older and wiser brother.

Freer's charity commenced on a very small scale and initially provided the poor with expensive curative apparatus. In the hospital's first month, staff saw just twenty-five cases, while fifteen other patients had been turned away.[23] In general, governors justified all cases, if not the very existence of the charity, by adhering to their main goal, which was a desire to transform disabled children into productive citizens. The decision to concentrate on the care of young patients was in fact very logical, given that it was possible to treat children's softer bones with splints and without resorting to more extreme surgical methods. Though he was a surgeon, Robert Chessher never attempted such radical methods of cure before his death in 1831.[24] In an age before antisepsis, the performance of more invasive techniques would have been highly dangerous, if not deadly, and, with time, undertaking such risky procedures would certainly have ensured the charity's failure. Instead, the main challenge facing its organisers was the need to raise funds sufficient to pay for the surgical instruments staff used to treat their

[19] Austin, *Robert Chessher*, 22.

[20] E. Harrison, *Pathological and Practical Observations on Spinal Diseases* (London, 1827).

[21] G. Eliot, *Middlemarch* (London, 1994), 434.

[22] Austin, *Robert Chessher*, 9.

[23] White, *Years of Caring*, 16.

[24] LeVay, *History of Orthopaedics*, 88.

young patients. In particular, Freer spent most of the charity's money on trusses: herniary complaints were prevalent locally.[25] Few operations were undertaken by the hospital's staff until the last quarter of the nineteenth century, when the work of Louis Pasteur and Joseph Lister, among others, transformed surgical procedures altogether. As a result, just as the charity's founder favoured compression in the treatment of aneurism, thereby rendering the knife unnecessary, so, too, was his approach to orthopaedics characterised by conservative methods.[26]

The hospital's work load does not appear to have increased very rapidly in its first eighteen months, a fact that would undoubtedly have suited Freer, who by this time enjoyed a very healthy private practice. Between summer 1817 and December 1818, staff treated 181 cases of rupture, or hernia, and eighty-seven body deformities, making an average of fifteen patients a month. With subscriptions appearing as rare as the governing committee's meetings, much of the charity's meagre income was spent on surgical instruments, primarily trusses. In fact, by the time George Freer resigned in the winter of 1820, given the number and demand of his private patients, activities at the hospital had been virtually suspended. The charity might even have ceased altogether had he not been succeeded by one of his former pupils, Martin Noble Shipton, who served the institution for nine years, even running it out of his home, before being joined in practice by Thomas Freer in 1825.

Though documents relating to the institution's earliest years are extremely scarce, some evidence relating to the hospital's first patients has survived in the publications of its medical staff. For example, some descriptions of the charity's work are contained in the sole book written by the hospital's first physician, John Darwall, another of Freer's pupils.[27] Perhaps most interesting, Darwall's work, *Plain Instructions for the Management of Infants* (1830), published three years before his untimely death, contains a chapter on 'deformities'. As one might expect, given his two-year involvement in the charity's work, Darwall commences this section on an optimistic note, suggesting many deformities encountered by medical practitioners in these years were treatable. While he is least optimistic about clubfoot, an ailment for which instruments offered little relief, he claims much more could be accomplished through rolling the deformed limbs of children, which, given their pliability at the earliest stages of life, could

[25] Austin, *Robert Chessher*, 48–9.

[26] Bunce, *Birmingham General Hospital*, 45.

[27] J. Reinarz and A. Williams, 'John Darwall, MD (1796–1833): the short yet productive life of a Birmingham practitioner', *Journal of Medical Biography* 13 (2005), 150.

be 'twisted into shape'.[28] By no means suggesting parents attempt to heal their own children, he further advised that any rolling be managed by surgeons, such as Thomas Freer, with their superior knowledge of anatomy. Additionally, he discouraged parents from forcing their infants to walk, as this, he claimed, often led to some of the deformities he was called on to treat in these years. So, too, did he deter guardians from sending children to work, Darwall having discouraged such practices until at least the age of five.[29] Only in very rare cases, most notably harelip, was he tempted to operate on children. Guided by Astley Cooper, 'consummate surgical practitioner, patron and kingmaker of the early decades of the nineteenth century',[30] whose lectures he undoubtedly attended while a student in London, Darwall, like Freer, was generally against such early operations.[31] Most practitioners similarly insisted on aiding the body's natural mechanisms of repair and not replacing them.[32]

Unlike many other specialisms, bone and joint diseases were not the preserve of quacks and charlatans. While Darwall's text discouraged parents from treating their children, it said almost nothing about unqualified practitioners. Primarily, this was due to the absence of quick and profitable ways to cure spinal curvature, for example. Though bonesetters treated orthopaedic disorders and in some cases established successful dynasties, including the Matthews of Oxfordshire and the Thomas family of Liverpool,[33] few appear to have advertised their services in Birmingham during these years. In general, their methods of manipulation and massage to treat 'cold' orthopaedic cases, such as congenital deformities, skeletal tuberculosis and tumours, had already been taken up by a new generation of qualified practitioners.[34] As a result, Birmingham's earliest orthopaedic surgeons, including Freer and Darwall, now looked to the work of London surgeons, such as Percival Pott and John Hunter, in order to treat fractures and other results of trauma more effectively.[35] While these were common enough in this industrial region, inevitably the majority of such accidents

---

[28] J. Darwall, *Plain Instructions for the Management of Infants, with Practical Observations on the Disorders Incident to Children* (London, 1830), 26.

[29] Ibid., xv.

[30] R. C. Maulitz, *Morbid Appearances: The Anatomy of Pathology in the Early Nineteenth Century* (Cambridge, 1987), 183.

[31] Darwall, *Plain Instructions*, 28.

[32] LeVay, *History of Orthopaedics*, 90.

[33] Lane, *Social History of Medicine*, 51.

[34] LeVay, *History of Orthopaedics*, 74.

[35] See, for example, P. Pott, *Some Few General Remarks on Fractures and Dislocations* (London, 1796).

were conveyed to the town's General Hospital, where patients more often endured amputation, rather than rehabilitation.

In general, the medical charities founded in these years were characterised by their moderate rates of growth. It was the unhurried and careful growth of the General Hospital, in fact, that essentially facilitated the emergence of the town's other medical charities. In turn, these initially grew more slowly than the town's primary medical institution, most subsequent medical charities commencing their lives as dispensaries. Governors appeared hesitant to expand operations in their first years, as fiscal prudence, along with the odd bit of medical work, seemed the perfect formula for attracting the public's enduring support. At least that was the theory. Despite much good work, by 1855, the Orthopaedic Hospital's annual income was still only £40, its subscribers numbering only twenty local families; in the same year, the General Hospital managed to collect approximately £7,000.[36] In an attempt to alter its fortune, by 1858, the small specialist hospital had moved to 21 Great Charles Street and the charity was reinaugurated in an effort to stimulate additional financial contributions. Though its board considered creating inpatient facilities, members agreed after reflecting on its difficult first decades that the charity could carry out its work more cheaply and efficiently as a dispensary. Pursuing a similar policy, the General's staff treated more than 15,000 patients annually, while admitting no more than 1,500 inpatients.[37]

Nevertheless, only at this stage did the charity really begin to resemble other general hospitals. Besides a consulting surgeon and physician, George Evans having replaced Dr Lloyd, who had succeeded Darwall, medical staff consisted of two honorary surgeons, Walter Freer and Charles Warden, who dispensed 'gratuitous advice from the premises' every Monday and Thursday at 2 pm. Those unable to walk were attended at home. The only person travelling greater distances than the medical officers was a full-time collector, who was also appointed shortly after the charity's move to Great Charles Street. That this gentleman travelled beyond the immediate vicinity of the hospital is suggested by a decision to include 'Midland' in the name of the institution at this time.[38] An annual report from 1874 states that more than a third of patients came from 'the surrounding districts'. Nevertheless, unlike the General, the Orthopaedic Hospital still treated only outpatients, inpatients being first admitted in 1877, when the charity moved to 81 Newhall Street. However, unlike orthopaedists in Chessher's time, the charity's medical staff was regularly undertaking corrective

---

[36] BCLA, Orthopaedic Hospital, Annual Report, 1875–6, HC/RO/Box 14.

[37] BCLA, GBH 416.

[38] BCLA, HC/RO/Box 14.

surgery, more than half of all cases of clubfoot at this time having been operated upon.[39]

A CHARITABLE VISION:
THE EYE HOSPITAL

Six years after the Orthopaedic Hospital's original launch in 1817, a third voluntary hospital was founded in Birmingham. On this occasion in 1823, a house at 35 Cannon Street was acquired by Samuel Galton, Joseph Hodgson and Dr Gabriel DeLys for the purposes of opening an eye infirmary in a move that represented the growing respectability of what had formerly been a quack specialty. In previous decades, the treatment of eye afflictions was in great measure confined to oculists, who very frequently acquired, with great notoriety, considerable wealth and importance, at least in public, if not medical, circles.[40] In contrast to the 'proud quacksalving mountbankes' who had previously and exclusively treated the town's visually impaired, the charity's founders emphasised both the importance of their venture and the limits of specialisation at the time. By no means confining themselves exclusively to the eye, specialists, like Hodgson and DeLys in Birmingham, as elsewhere, emphasised that the eye was not detached from the bodily system, the textures of the eye often sympathising with the state and diseases of the general constitution.[41] For example, it was claimed that the power of the retina in women was infrequently diminished during lactation or as a consequence of amenorrhea.[42] The proper treatment of the eye, therefore, demanded a comprehensive knowledge of the body. As a result, such ailments could no longer be locally treated with washes, drops and ointments, as were formerly dispensed by legions of untrained quacks.

Rather than initiating this professional struggle over the eye, Birmingham doctors were merely following trends emerging elsewhere. Already two decades before the foundation of Birmingham's newest medical charity, in 1805, John Cunningham Saunders, the former anatomy demonstrator, upon the advice of Astley Cooper, established the London eye dispensary (which would later be renamed Moorfields).[43] Then, between 1808 and 1832, a further nineteen eye infirmaries, Birmingham included,

39 Austin, *Robert Chessher*, 50.

40 P. Trevor-Roper, 'Chevalier Taylor – Ophthalmiater Royal (1703–1772)', *Documenta Ophthalmologica* 71 (1989), 113–22; L. Davidson, '"Identities ascertained": British ophthalmology in the first half of the nineteenth century', *Social History of Medicine* 9, 3 (1996), 318.

41 Davidson, 'Identities ascertained', 331–3.

42 R. Middlemore, *A Treatise on the Diseases of the Eye and its Appendages*, vol. 1 (London, 1835), 31.

43 Davidson, 'Identities ascertained', 320.

were established, primarily in the south-west and the industrial north, by an English public that valued its sight above all other senses.[44] Much like eye institutions, publications on the anatomy and afflictions of the eye had also multiplied by 1830. The foundation of eye hospitals also coincided with philanthropic interest in the blind that began in France in the 1780s and simultaneously resulted in the establishment of a large number of schools for the blind in England.[45] While a national trend is apparent, locally the establishment of such an institution had much to do with an earlier example set by George Freer, the founder of the Orthopaedic Hospital. As one of the surgeon's apprentices, if not his star pupil, having illustrated his master's work on aneurism, Joseph Hodgson would have grasped the way in which medical institution-building could noticeably accelerate the career of a young practitioner. As has been argued elsewhere, 'it was in the assistance that the hospitals provided to the upwardly mobile doctor that the key to their foundation can be found'.[46] The Eye Hospital, the tenth institution of its kind in England, is perhaps the best early example of this phenomenon in Birmingham.

While the professional disease of blocked promotion helped create many hospitals throughout England in the nineteenth century, epidemics or a particular concentration of ophthalmic injuries also encouraged the establishment of the nation's first eye hospitals. For example, the foundation of these ocular charities is generally seen as a reaction to outbreaks of conjunctivitis, or Egyptian ophthalmia, as this disease was called when it crippled many British troops in Egypt in the late eighteenth century,[47] despite earlier appearances among regiments stationed in Sicily, Malta and Gibraltar. Though not established in response to a wide outbreak of ophthalmia, the Birmingham eye charity was, without a doubt, better suited to the needs of the regional economy than was the Orthopaedic Hospital, eye injuries having become prevalent in the town's numerous workshops, not to mention the mines of the adjoining Black Country, over the preceding century; despite the proliferation of industrial hazards to sight, small-pox remained one of the major causes of blindness among the inmates of

---

[44] A. Sorsby, 'Nineteenth century provincial eye hospitals', *British Journal of Ophthalmology* 30 (1946), 504; Davidson, 'Identities ascertained', 325–6.

[45] Davidson, 'Identities ascertained', 323.

[46] L. Granshaw, '"Fame and fortune by means of bricks and mortar": the medical profession and specialist hospitals in Britain, 1800–1948', in *The Hospital in History*, ed. L. Granshaw and R. Porter (London, 1989), 200.

[47] H. Corlett, '"No small uncertainty": eye treatments in eighteenth-century England and France', *Medical History* 42 (1998), 220; Davidson, 'Identities ascertained', 315–17.

English blind institutions well into the nineteenth century.[48] In 1780, for example, staff at the General Hospital treated twenty-three patients with ophthalmic complaints, including twelve cases of 'sore eyes', three cataract and two cases of ophthalmia.[49] Subsequent years saw approximately two dozen comparable cases attend the institution annually. A concentration of eye injuries in the local mining industry had similarly encouraged administrators of the first Cornish hospital to establish an ophthalmic ward.[50] In much the same way, from its foundation, the Eye Hospital in Birmingham served the region's industrial base and especially the metal and glass trades. Moreover, the regional appeal of all specialty institutions was apparent from their first appearance, hospital administrators in Birmingham, those of the Eye Hospital included, having frequently added 'Midland' to their names better to represent the communities they served. However, the Eye Hospital seems to have appealed more widely than all other specialist hospitals in these years, its patient registers and subscription lists recording individuals, and particularly businesses, from a 20-mile catchment area, beginning with many local collieries.[51]

As one might therefore expect, the charity grew more quickly than the town's other small voluntary hospitals. In 1824, the same year the General's staff treated 4,000 patients, the Eye Hospital was already attending 1,733 cases annually, albeit on a much smaller budget, similar figures having been reported by eye institutions in Manchester and Liverpool at this time.[52] Instruction at the hospital, as at Manchester, also commenced very quickly. By 1829, a year after he was appointed a surgeon to the charity, a young and energetic Richard Middlemore announced his intention to deliver a series of lectures on the diseases of the eye, a practice he insisted would have received little encouragement a generation earlier, or even have been prohibited by hospital governors;[53] similar lectures were first delivered in Manchester a decade earlier in order to raise funds for the town's eye institution.[54] In 1835, the year in which Middlemore published his twenty collected lectures on the diseases of the eye, the Birmingham charity is said to have afforded relief to some 16,000 of the town's poorest, visually impaired

---

[48] Corlett, 'No small uncertainty', 233.

[49] BCLA, General Hospital, Birmingham, Patient Register, 1779–88, GH/4/2/560.

[50] C. T. Andrews, *The First Cornish Hospital* (Penzance, 1975), 43.

[51] Birmingham Central Library Local Studies (BCLLS), Eye Hospital, Annual Report, 1861.

[52] BCLA, Eye Hospital, Committee Minutes, 1823–57, MS 1919; F. S. Stancliffe, *The Manchester Royal Eye Hospital, 1814–1964* (Manchester, 1964), 16–17.

[53] Middlemore, *Treatise on the Diseases of the Eye*, 1.

[54] Ibid., 14; Stancliffe, *Manchester Royal Eye Hospital*, 17.

inhabitants.[55] However, still sharing more of the Orthopaedic Hospital's characteristics, its 'humble' and 'unimposing' premises in Cannon Street were first rebuilt in 1838, the same year the governors of Manchester Eye Hospital also began to discuss a purpose-built ophthalmic institution.[56]

As one might expect, Middlemore's published lectures shed considerable light on the activities of Birmingham's eye charity in these years. Much like other texts published in these decades, Middlemore's work is a particularly detailed description of eye diseases, the first volume running to 800 pages. The size of the textbook alone appears to suggest that ophthalmology was no longer a narrow speciality, but a complicated and comprehensive medical field in itself. Despite its size, the book discusses only a few improvements which had very lately been introduced in the treatment of ophthalmic diseases. For example, Middlemore considers the use of strychnia in the treatment of partial blindness and ptosis (p. 23), as well as the use of silver nitrates to cure many acute and chronic diseases of the conjunctiva (p. 55). The employment of turpentine and iodine was also recommended in some cases of inflammation, many of which had previously been treated with mercury.[57]

Perhaps more interestingly, Middlemore also refers to particular trades that excited the charity's common ailments, such as conjunctivitis. Tradesmen found to be particularly susceptible to this condition were tailors, shoe-makers, engravers, working-jewellers and gilt-toy makers, and especially those working by artificial light.[58] Working from his case notes, one is also able to reconstruct the sort of treatment some of the town's less fortunate engravers and enamellers might have experienced had they been admitted to the charity with, for example, severe conjunctivitis. Generally, Middlemore's remedies were premised on beliefs prevailing in the early nineteenth century that diseases were overstimulating – and therefore characterised by inflammation – and best treated by heroic depletive therapy.[59] Additionally, a detailed discussion of Middlemore's methods of treatment demonstrates the way in which his therapies differed from the locally applied remedies of the quacks he sought to replace.

When faced with the average strong, healthy person, Middlemore's treatment would have commenced with some generous blood-letting, perhaps 20 to 60 ounces, or until the patient showed signs of faintness.

---

[55] Middlemore, *Treatise on the Diseases of the* Eye, 14.

[56] Stancliffe, *Manchester Royal Eye Hospital*, 28.

[57] Middlemore, *Treatise on the Diseases of the Eye*, 24.

[58] Ibid., 50.

[59] J. H. Warner, *The Therapeutic Perspective: Medical Practice, Knowledge, and Identity in America, 1820–1885* (Princeton, NJ, 1997), 91–2.

He might then have applied a blister behind the ear to counter the inflammation observed in the labourer's eye, cupping glasses having not surprisingly been purchased by the charity in these years. In cases where inflammation spread to both eyes, a blister would be placed behind the neck of the afflicted individual.[60] In either case, eyes would be bathed with warm water or milk, or, if patients appeared to be in considerable pain, a strong aqueous solution of opium, a decoction of poppies or a fomentation of hops would be substituted.[61] In general, the surgeon endeavoured to prostrate the powers of the system, hence his readiness to bleed patients and administer some active purgative, usually calomel, or an emetic, such as tartarised antimony.[62] Patients would also be placed on a reduced or antiphlogistic diet, which denied them meat and stimulants of any kind. Over the following several days, bleeding would continue through the application of numerous leeches to the head, Middlemore suggesting these therapeutic worms gradually be placed nearer the seat of the disease as treatment progressed, though usually avoiding the eye-lid. Washing of the eye would continue, only now nitrate of silver drops were applied by means of a soft piece of sponge or linen.[63] Any exercise of the eyes was to be discouraged, the patient in most cases of inflammation being made to wear a green shade, or a square bit of linen tied loosely around the head. In the most fortunate cases, the remainder of a patient's recovery period was to be spent in a darkened room.

In some cases, it was believed that discharge from the eyes was the result of suppressed discharges elsewhere, and therefore more active intervention was called for. For example, it was frequently observed that a purulent eye disease commenced only after discharge from another organ or orifice had ceased. In such cases, the eye charity's surgeons may have attempted to restore the absent or deficient discharge by instituting some artificial drain or issue, such as a deep stitch, from the affected part. Occasionally, these issues, otherwise known as setons, were permitted to remain open for years. Regardless of the treatment advocated, doctors' case notes testify that most patients suffered for prolonged periods from their ocular ailments. According to Middlemore, few poor patients received treatment in the early stages of their illnesses, most having approached the charity only after attempting to cure themselves with traditional folk remedies. This often included rubbing the eye with spittle, applying a poultice, blister, or even

---

[60] Middlemore, *Treatise on the Diseases of the Eye*, 55.
[61] Ibid.
[62] Ibid., 74.
[63] Ibid., 84.

bathing the organ in tea or urine.[64] Even in the most straight-forward of cases, considerable injury was inflicted when sufferers or their friends attempted to remove objects from eyes, often mistaking 'stains of the cornea' or other birthmarks for the offending wood chip, or metal filing.[65]

While spending much of his time attempting to discourage such practices amongst the poor in Birmingham, Middlemore also hoped to discourage certain practices among fellow practitioners. For example, when treating inflammation, he encouraged doctors to bleed their patients from the arm, or by applying leeches behind their ears, not the nose or anus as had been formerly in vogue. Neither did he encourage surgeons to open the temporal artery as this vessel was difficult to restrain, often requiring tight bandages which compressed the surrounding vessels and heated the head. Additionally, he advocated that practitioners examine the eye by carefully lifting the lid, taking firm hold of a few eyelashes and not apply rough and undue pressure on the eye as he had so often seen, especially in rural practices. Middlemore also believed the time had come to discard many strange local applications, such as hen's dung, bags of mallow and roasted apple poultices.[66] Though still encouraging scarification of inflamed tissues with a view to lessening their vascularity, he suggested this be done with a needle or lancet, not a common thistle, a steel rasp, or a 'barley beard' as was formerly a favourite practice.[67] Neither did he favour the application of spirits or vinegar, nor the blowing of snuff into the eye, preferring instead the application of olive oil, especially for burns of the conjunctiva. He equally discouraged puncturing the cornea in order to relieve the pain and tension that generally accompanied severe forms of acute inflammation of the eye, the most common affliction seen at early eye hospitals.

Though drawing on much evidence that appeared to confirm the contagious nature of numerous eye infections, Middlemore's text nevertheless represents a transitional, rather than a revolutionary period in the history of ophthalmology. For example, though discouraging parents from using the linen or sponge with which they washed the pus from their children's eyes for any other purposes, he remained unconvinced that ophthalmic diseases were caused by contagion alone. Neither was he able to discount the concept of spontaneous generation entirely. As evidence, he draws on several accidents and cases of self-experimentation that had been brought to his attention over the years. One involved a gentleman named Machesy,

[64] Ibid., 56–7.
[65] Ibid., 543.
[66] Ibid., 83.
[67] Ibid., 71.

who is said to have applied a piece of linen soaked in purulent discharge
to his eyes while residing with troops in Egypt and experienced little more
than a short period of discomfort. Another involved nurses and medical
students at the Birmingham Eye Hospital whom he had observed coming
into contact with ocular discharge when syringing the eyes of adult patients
suffering from similar diseases, with no ill consequences. More unusually,
he relates the case of a hospital assistant who applied gonorrhoeal matter
from his own urethra to the conjuntiva of his eye, without producing the
disease in that organ.[68] Faced with such testimony, Middlemore became
convinced that contagion alone did not produce inflammation of the eye,
the process being aided, he suspected, by a variety of circumstances, including want of cleanliness, disordered health, exposure to brilliant sun, or the
peculiar condition of the atmosphere. This did not prevent him, however,
from suggesting that the linen used by those affected with ophthalmia
should be strictly set apart for their sole use and the usual precautions preventing the improper transfer of the contagious fluid be employed, much to
the benefit of the hospital's first patients.

While medical staff continued to struggle with ideas of disease trans-
mission, the hospital's administrators had more success with the financial
side of their project in their first decades. In 1846, after some years in poorly
maintained premises, the charity's managing board, like their colleagues in
Manchester some eight years earlier, even began to discuss a permanent,
purpose-built hospital.[69] Two years later, a building fund was opened and
attracted sufficient support to permit the purchase of a failed local poly-
technic institution for £1,900. Unwilling to contemplate similar failure, in
1849, staff opened a fifteen-bed eye hospital in Steelhouse Lane to house
the charity's first inpatients. This move would only further transform the
way in which this particular specialty was perceived locally. By the middle of the nineteenth century, local residents had already learned to deal
with ocular complaints very differently than in past decades. The fact that
Birmingham's visually impaired inhabitants were no longer amongst the
largest groups targeted by quacks in local broadsheets would have pleased
early ophthalmic specialists, such as Richard Middlemore. For example, of more than 260 medical advertisements appearing in *Aris's Gazette*
between 1840 and 1870, less than half a dozen peddled miracle cures for
eye ailments.[70] In contrast, the hospital advertised its own meetings in

---

[68] Ibid., 189.

[69] BCLA, Eye Hospital, Committee Minutes, 1823–57, MS 1919

[70] *Aris's Gazette*, 1 January 1840 – 30 December 1870. Many thanks to Andrew
Edwards for his analysis of unorthodox medicines advertised in *Aris's Gazette* in
these years.

newspapers from Worcester, Stafford, Derby and Coventry.[71] Despite these changes, a far greater revolution in ophthalmology was about to commence.

Within a few years of its appearance, the ophthalmoscope was already regarded as the most important technological invention within ophthalmology. According to an anonymous medical personality writing in Birmingham in 1863, 'the ophthalmoscope is to the ophthalmic surgeon what the telescope is to the astronomer or the microscope to the naturalist'.[72] Invented by the physiologist Hermann von Helmholtz (1831–94) in 1851, the Prussian professor's 'eye-speculum' allowed practitioners direct access to the eye. As such, eye specialists could now examine the living retina and link symptoms with pathological signs. Although the term 'ophthalmoscope' did not appear until September 1853, the instrument equally transformed the literature of ophthalmology, changing as it did the way in which pathology and disease was viewed. Not surprisingly, the the Royal London Ophthalmic Hospital reports, the first publication dedicated exclusively to the subject of ophthalmology, appeared four years later. That same year, 1857, the first international conference for ophthalmology was held in Brussels, two decades before a similar general international medical conference would be organised.

Similarly, while the early nineteenth century saw ophthalmology principally practised by surgeons,[73] this changed with the advent of the ophthalmoscope. Ophthalmology had become a specialty that influenced and overlapped with other areas of medicine and surgery, including neurology, as would be demonstrated by the Moorfields physician, John Hughlings Jackson. Not the easiest instrument to learn, the ophthalmoscope would soon be incorporated into the clinical instruction of all medical students, and the eye hospital appeared the logical site from which to diffuse this knowledge. Despite moving into a building formally dedicated to educational functions, however, the Birmingham Eye Hospital's position as a teaching hospital was late to develop. Systematic attempts at instructing medical practitioners in Birmingham commenced in premises far less suited to such grand educational ventures, and it is on this significant development than the next chapter focuses.

[71] BCLA, Eye Hospital, Committee Minutes, 1857–66, MS 1920.

[72] Scutator, *The Medical Charities of Birmingham: Being Letters on Hospital Administration and Management* (Birmingham, 1863), 49.

[73] Davidson, 'Identities ascertained', 317.

CHAPTER 3

Birmingham School of Medicine and the First Provincial Teaching Hospital

APPEARING IN THE PROVINCES almost as frequently as voluntary hospitals during these years were medical schools. Primarily, this was the result of a decision by the Royal College of Surgeons in 1826 to recognise courses offered by provincial medical instructors, as long as these individuals possessed the membership of the college and satisfied their teaching requirements.[1] Just as private anatomy schools had flourished in eighteenth-century London to supplement the education of hospital surgeons' apprentices, a number of enterprising practitioners commenced similar anatomical instruction in various provincial centres. The first to offer a course of anatomical instruction in Birmingham was the surgeon Thomas Tomlinson, whose lectures and other medical views were published in 1769 as *Medical Miscellany*.[2] Numerous others residing in the English provinces organised similar classes. For many it was a good way to fill the awkward gap between qualifying and securing an adequate medical practice. Classes always appear to have attracted about a dozen students and usually some local practitioners interested in the ways medicine had changed since qualifying. While some lecture series were short lived, others developed into proper schools offering more than a few practical lessons in anatomy. Outside Oxford and Cambridge, the first three provincial medical schools were founded in Manchester, Sheffield and Birmingham.[3]

LEARNING TO HEAL:
THE EARLY YEARS OF A PROVINCIAL MEDICAL SCHOOL

Like the orthopaedic and eye charities, medical instruction in Birmingham commenced on a small scale. The first school, in fact, was located in a house at 24 Temple Row, the building formerly occupied by the town's first dispensary, which had become the home and medical practice of Edward Townsend Cox and his son, William.[4] Launched in 1825, instruction

---

[1] Z. Cope, *The Royal College of Surgeons of England* (London, 1959), 46.

[2] T. Tomlinson, *Medical Miscellany* (London, 1769, republished 1774).

[3] S. T. Anning, 'Provincial medical schools in the nineteenth century', in *The Evolution of Medical Education in Britain*, ed. F. N. L. Poynter (London, 1966), 121.

[4] J. T. J. Morrison, *William Sands Cox and the Birmingham Medical School* (Birmingham, 1926).

included primarily anatomy lectures and, more valuably, the dissection of human cadavers, access to which was certainly facilitated by the work that Edward Cox undertook as one of the surgeons to the workhouse infirmary. The appeal of these practical anatomy courses would continue well beyond the passage of the Anatomy Act in 1832, which was intended to increase the supply of cadavers for the purposes of medical education and put an end to the disreputable practice of body-snatching.[5] The work of many influential French clinicians of the early nineteenth century, who traced diseases to their corresponding lesions in bodily tissues and encouraged their students to dissect, only strengthened localistic and solidistic tendencies and bolstered interest in pathological anatomy.[6] Consequently, at a time when disease was being localised, medical education too had begun to root more firmly in local, or provincial, settings.

Supported by the town's senior practitioners and having only just completed his own medical education in London and Paris, William Sands Cox and his lessons had been popular since they commenced, leading the educational undertaking to expand more rapidly than the town's smaller medical charities. In this way, his courses were much like those of Edward Grainger, whose Webb Street anatomy school, one of the twelve anatomical schools in London at this time,[7] was an inspiration to Cox. Furthermore, Cox lived with the anatomy instructor while undertaking his medical studies in the capital.[8] Unlike Grainger, who did not teach in his native Birmingham, Cox, again aided by several senior local practitioners, soon attempted to offer a more complete course of medical studies, not simply anatomical instruction. While William Sands Cox taught anatomy, physiology and pathology, all other lectureships were offered to the surgeons and physicians at the General Hospital and Dispensary in order of seniority. As a result, he was joined by two physicians to the General Hospital, Richard Pearson and John Kaye Booth. The former, besides delivering the institution's first inaugural address, initially taught materia medica and medical botany, the latter becoming responsible for the principles and practice of physic. One of the hospital's surgeons, Alfred Jukes, was the first instructor in surgery, while John Woolrich taught chemistry and pharmacy.[9] John Ingleby, on the other hand, taught midwifery and the diseases of women

5   R. Richardson, *Death, Dissection, and the Destitute* (Chicago, 2001).

6   E. Ackerknecht, *Medicine at the Paris Hospital, 1794–1848* (Baltimore, 1967), 3–58.

7   R. C. Maulitz, *Morbid Appearances: The Anatomy of Pathology in the Early Nineteenth Century* (Cambridge, 1987), 178.

8   W. H. McMenemey, 'William Sands Cox and the stoicism of Elizabeth Powis', *Medical History* 2 (1958), 113.

9   University of Birmingham Special Collections (UBSC), MS, Minute Book, 1831–8.

and children, the former subject gaining somewhat in importance in 1852 when a diploma in midwifery was instituted by the Royal College of Surgeons, every candidate for the membership of the College having been required to attend two courses of lectures on 'the obstetric art and science' since 1828.[10] While the latter were taught by Ingleby, the local man-mid-wife did not live to witness the diploma's introduction, as he died of gout in 1845 at the age of fifty-two. Nevertheless, he ensured the centrality of his subject to medical education locally by leaving £2,000 to endow a lecture series at the school on advancements in obstetric medicine and surgery.

Though Cox's lectures were the most thorough experiment in medical education in Birmingham when lessons commenced in 1825, they were not the first to have been offered in the Midlands. In Birmingham, that distinction went to Thomas Tomlinson, who, in 1769, a year after William and John Hunter had commenced anatomical instruction at their Great Windmill Street school in London, delivered a series of twenty-eight lectures to local medical apprentices.[11] Despite such educational opportunities, most medical students continued to learn by way of apprenticeship, and clinical material was therefore limited to those cases that came before their masters. As such, the establishment of a general hospital in Birmingham was a significant event in the history of medical education locally. From its establishment in 1779, the doors to the General Hospital, much like those founded half a century earlier in the capital, would have been open to pupils of its medical staff. Though only attendance at a London hospital would count towards medical students' qualifications until the end of the third decade of the nineteenth century, this would not have deterred many from gaining experience in provincial wards and dispensaries.

Though most young practitioners sought membership of the College of Surgeons, its council was slow to draw up a syllabus or requirements for candidates who desired to sit their professional examinations. As a result, the Society of Apothecaries led the way in 1815, when their recommendations 'for better regulating the Practice of Apothecaries throughout England and Wales' were introduced. By creating their licentiate, the organisation not only raised the status of the profession by creating a recognised qualification, but encouraged further developments in provincial medical education. Significantly, in 1817, staff at the General Hospital in Birmingham made their first official efforts to develop the educational functions of their charity. In that year, the hospital's medical board offered to set up a museum, as well as a private medical and surgical library and promised to support this initiative provided an appropriate portion of the governing committee's

---

[10] Cope, *Royal College of Surgeons*, 130–1.
[11] Tomlinson, *Medical Miscellany*, v–vi.

funds were devoted to the same facilities.[12] Most medical students, like the senior pupil indentured to Mr Jukes's in 1818, however, would continue to travel to London for a period of six months in order to attend anatomical and surgical lectures, walk the wards of metropolitan hospitals and generally satisfy the educational requirements of the Royal College of Surgeons. Surprisingly, the ledgers of the General Hospital in Birmingham contain no further mention of pupils until 1823, when Joseph Hodgson reported to the governors that pupils' drawers in the hospital surgery had been broken into and many articles stolen.[13] A decision by the general committee in 1826 to place beds for surgeon's pupils in the house suggests this would not be their final appearance in hospital records.[14]

Like those of his predecessors, William Sands Cox's course of lectures in 1825 would primarily have provided students with a better understanding of the human body. They would also have encouraged those in attendance to undertake more minute examinations of organs and tissues and so transform the medicine of symptoms as taught by a previous generation of practitioners into a medicine of lesions, as was instructed in schools in London and Paris at this time.[15] While students learned the importance of the local, so too did the parents and guardians of young medical prospects. Sands Cox, for example, stressed the fact that his locally held sessions would 'shield his students from [the] coarse influences to which they would be exposed in the metropolis'.[16] Most other provincial medical schools similarly stressed the moral safety of a local education. Equally important to attracting students was that a provincial medical education was cheaper than sending a son to London.

When more students began to attend his lectures in 1828 than his Temple Row practice could accommodate, Cox acquired a piece of land in the centre of Birmingham at Snow Hill. On this site, which previously housed a warehouse and carpenter's shop and yard, he commenced to construct a school, comprising lecture theatre, museum and library at a cost of £1,000. While access to clinical material on the wards of the General Hospital and workhouse infirmary was ensured by Edward Johnstone's and Edward Townsend Cox's links to these institutions, it was in these years that the school more generally began to resemble an educational institution. To begin with, a school council and various committees were established. In addition, a minute book was kept from 1831 by a general management

---

[12] BCLA, GH/1/2/6.

[13] BCLA, GH/1/2/7.

[14] BCLA, GH/1/2/8.

[15] Ackerknecht, *Medicine at the Paris Hospital*, 3–58.

[16] Morrison, *William Sands Cox*, 20–1.

committee that met weekly at the busiest of times and less frequently otherwise. More regular were certain institutional rituals, such as annual dinners, so characteristic of all medical schools at the time, given their promotional value. An annual dinner in 1832 held at Dee's Royal Hotel was just one such well-publicised occasion. Besides being attended by more than sixty guests, most of whom were students, news of the evening and its endless series of toasts were enthusiastically reported in local newspapers. The event was reported much less favourably by the *Lancet*'s editors, who suggested that 'the less puffing that takes place in the shape of anniversary festivals, the better'.[17]

It is clear from surviving documents that, despite such grand gestures, the school was still in its earliest stages and developing daily. To begin with, the curriculum was expanding each term, with the local coroner and 'friend of the poor' Birt Davies commencing a series of lectures on the subject of forensic medicine in the summer of 1832;[18] a course in forensic medicine was made a prerequisite for the Society of Apothecaries exam in 1830 and, as a result, were rapidly organised by most provincial schools. That same season, anatomical demonstrations, as distinct from anatomy lectures, were also commenced under the direction of George Elkington in order to ensure that schools conformed to the latest requirements of the Royal College of Surgeons. Not surprisingly, Cox was more hesitant about offering those courses that were not yet compulsory, such as the classical instruction suggested by one of the school's governors that same year. With each new course came the added expense of both books and, in many cases, museum specimens which lecturers used to illustrate their lessons. To the school's advantage, the departure of each aged instructor, often preceded a bequest of valuable teaching materials. For example, when James Johnstone assumed the chair of Materia Medica only just vacated by Richard Pearson in 1832, the school also received many of the learned scholar's books and botanical preparations. Shortly afterwards, the inheritance was augmented by the donation of a complete collection of medicinal plants from the Chelsea Gardens.[19] This generous offer was overshadowed only by legislation that granted all unclaimed medical bodies to medical schools, the result of the Anatomy Act passed that same year.

By 1834, the school again strategically relocated, taking larger premises across from the recently constructed Town Hall for £100 annually.[20]

---

[17] *Lancet*, 12 September 1835.

[18] *Lancet*, 25 May 1839.

[19] UBSC, MS, Minute Book, 1831–8.

[20] Morrison, *William Sands Cox*, 33.

Its new buildings, formerly a place of worship,[21] were located on Paradise Street and had been acquired by Edward Townsend Cox at public auction for £1,500 in May 1834. Drawing on existing regulations from medical schools in London, rules were drawn up which clearly outlined the financial responsibilities of each lecturer. The first session in the school's new premises was opened with an inaugural lecture by John Johnstone, and the chief subject of admiration was the school's museum. According to evidence presented before the 1834 Select Committee on Medical Education, the Royal College of Surgeons did not recognise provincial schools unless in possession of such collections.[22] Not surprisingly, donations to the museum subsequently increased to a remarkable extent. In return, the school's governors opened the museum to the public free of expense three days a week, from noon until four, during August and September.[23] Interestingly, the museum was also open only three days a week to students during term time, despite its importance to their studies.

Besides the school's advertisements, which listed lecturers and their courses, the best indication of the institution's curriculum was the size of its lecturers' fees, which were based, as already mentioned, on the number of lectures delivered. By the summer of 1834, having delivered the greatest number of lectures, and in receipt of the largest proportion of fees, William Sands Cox still paid more towards the school's upkeep than any other instructor. However, he now paid only 9 guineas, having relinquished some teaching to medical practitioners who were equally qualified and could spare the required time, given his increasing administrative duties. Five other instructors paid 6 guineas, while George Knowles and Birt Davies, paid only three, given botany's and forensic medicine's subsidiary positions in the curriculum. Besides their teaching duties, three lecturers were also appointed each quarter to audit the school's accounts. Despite this additional burden, most departments continued to report much progress. The obstetric department was singled out in February 1835, for example, given many recent additions of teaching specimens to its collection, which now included 'two highly interesting specimens of disease of great practical value'.[24] While the lecturers were pleased to have finally acquired the uterus of a woman, a resident of Coventry in this case, the committee was equally proud to proclaim that the female from which it was 'torn' was in 'perfect health'.[25] The exact provenance of numerous other

[21] Hutton, *History of Birmingham*, 385.

[22] Select Committee on Medical Education, 1834, 1835.

[23] UBSC, MS, Minute Book, 1831–8.

[24] Ibid.

[25] Ibid.

specimens acquired in these and subsequent years is considerably harder to uncover.

In general, corpses were more readily available in Birmingham than at other schools since the passage of the Anatomy Act in 1832. Twenty-three alone were dissected by students in 1832, far more than were made available to schools in Bristol (17), Manchester (13) and Sheffield (11).[26] The following year, another twenty bodies were dissected at the school, more than were available to students at Sheffield (19), Leeds (9), Bristol (7), while those attending courses in Hull and Nottingham had none.[27] Lecturers at Cambridge acquired corpses with equal difficulty. In contrast, in 1835, Birmingham's anatomy lecturer reported that twelve bodies had been inspected during the winter session, the certificates of burial having been transmitted to the Home Office according to the regulations of the Anatomy Act. Furthermore, the school was said to pay 'every possible attention and respect to the feelings of the public, by conducting the rite and practice of sepulchre with solemnity and decorum'.[28] Additional thanks, not surprisingly, were also conveyed to the overseers for their liberal feelings and for the great aid they afforded the institution through the donation of unclaimed bodies.

The auditors' report that year set the school's quarterly costs at a little less than £165; given its detail, this provides some idea of the institution's early appearance.[29] For example, expenses include those for blinds in the museum and lecture room, a table for the library, a cast iron railing for the gallery in the museum, glass cases for the natural history collection, shelving for the same, and it even mentions a 'chemical room'. The complete property of the institution was insured at the Birmingham Fire Office for £800. The auditors, on the other hand, were adamant that only a small amount of the total expenditure could be 'justly appropriated to the Lecturers'.[30] Due to many difficulties in defining the exact proportions for which lecturers were responsible, the auditors suggested the whole sum should come from the institution's General Fund. In any case, costs would only continue to increase, especially as a subcommittee was formed half a year later to discuss the desirability of establishing lectureships in mathematics and natural philosophy as required by those seeking the qualifications of the Society of Apothecaries. Longer series of lectures were also

---

[26] R. Franklin, 'Medical Education and the Rise of the General Practitioner, 1760–1860' (PhD diss., University of Birmingham, 1950), 188.

[27] Anning, 'Provincial medical schools', 126.

[28] UBSC, MS, Minute Book, 1831–8.

[29] Ibid.

[30] Ibid.

required by instructors of other subjects. Some fiscal restraint was finally exercised by resolving that each new appointment should be 'in possession of the means requisite for the full illustration of his Lectures'.[31] Additional income was generated by selling tickets to the museum and library for a shilling through the principle booksellers in Birmingham. Surprisingly, the committee was more hesitant about appointing a lecturer in natural history, despite the existence of a natural history museum and the fact that most monthly donations to the school were illustrative of the same subject.[32]

Less often mentioned in the school's first records are its earliest students. The summer of 1835, however, was also unusual in terms of student intake, and therefore the subject was given greater attention in school records. The new session began with ninety students on the register, making the school the second largest in the provinces after Manchester.[33] This was an increase of more than forty on any other previous year, and twelve others were to have passed their examinations in London. Not surprisingly, like the patients they observed on their ward rounds, the majority of students came from neighbouring towns and villages, though a few came from as far away as Devonshire and Nottingham. Moreover, according to the minutes, since 1828, not a single student who had been educated at the school had been rejected by examiners of the Royal College of Surgeons.[34] Less publicised was that this amounted to only five candidates at a time when most of the largest schools in London could boast of more than 100 licentiates.[35] In general, most students trained only for short periods in Birmingham before moving to more prestigious schools in the capital or even Edinburgh, where they might also qualify as physicians.

In order to encourage additional success, the school had also recently begun to offer a number of prizes to its students. Though staff had awarded essay prizes in various subjects since 1833, incentives for students to excel increased noticeably by 1835. In that year, John Meredith was thanked for the gold medal that was to be presented to the best essay on a surgical subject, and the Revd Dr Arnold, Head Master of Rugby School, was thanked for his ten-guinea prize, awarded to the best essay 'On the influence of air and soil as affecting health';[36] interestingly, the latter prize is

---

[31] Ibid.

[32] J. Reinarz, 'The age of museum medicine: the rise and fall of the medical museum at Birmingham's Medical School', *Social History of Medicine* 18, 3 (2005), 427–8.

[33] Anning, 'Provincial medical schools', 126.

[34] UBSC, MS, Minute Book, 1831–8.

[35] *Report from the Select Committee on Medical Education*, 1834, 18 (of appendix).

[36] UBSC, MS, Minute Book, 1831–8.

also the first mention of instruction in public health at the school. In 1836, the school itself was the recipient of a far greater prize when it was granted royal patronage and became the Birmingham Royal School of Medicine and Surgery. While this alone might have improved enrolment, many new prizes were created that same year. Dr Booth donated 5 guineas, which were to be awarded to a pupil attending the surgical and medical practice of the General Hospital and presenting the best clinical reports on a certain number of cases. At the end of the same session, J. A. James offered 10 guineas for another essay with perhaps the most rambling of titles, namely 'On the influence upon health of alcoholic drinks as an article of diet including of course the consideration whether any quantity of any kind be necessary for the maintenance of health even in the case of those who are engaged in laborious occupations'. Another prize of equal value, but of a less-specified nature, was donated by James Upfill, to be presented on any subject the lecturers saw fit, whereupon Dr James Johnstone immediately proposed capillary vessels and, more specifically, their division, general conformation, situation, modes of communication, structure, physical and vital properties, functions and morbid anatomy. The first recipient of the prize was Thomas Clark Roden for his essay on the 'Valvular Structure of the Veins'; all subsequent prize essays in the following decade, save that of 1843, were deemed worthy of publication.[37] Influenced by practices at Oxford and Cambridge, the provincial schools appear to have developed a system of prizes and scholarships more quickly than educational institutions in London.[38]

The following year, additional donations, not to mention influence on the subjects taught at the school, came in the form of a book prize from Birt Davies, 21 guineas from Dr Jephson of Leamington Spa and another 10 guineas from Revd Josiah Allport for the best essay 'on animal, vegetable and chemical poisons', with a promise for 10 guineas more should the paper be judged worthy of publication. Finally, on the first day of 1838 the committee reported the donation of £1,000, the interest of which was to be applied for the life of the institution to two annual prize essays in order to combine religious with scientific study, or more specifically, 'to make medical and surgical students good Christians as well as able practitioners in medicine and surgery'. The donor therefore stressed the essay be of a religious as well as a scientific character, the student to acknowledge both 'evidences of facts and phenomena which anatomy, physiology and pathology so abundantly supply but always … set forth by instance

37 UBSC, Vaughan Thomas Collection, MSS 281/i/10.

38 K. Waddington, *Medical Education at St Bartholomew's Hospital, 1123–1995* (Woodbridge, 2003), 84–5.

or example the wisdom, the Power and Goodness of God as revealed and declared in Holy Writ'.[39] Unlike past prizes, which were judged by the lecturers alone, these were to be judged by the Dean of Lichfield Cathedral, Chancellor of the Diocese of Lichfield and Revd Vaughan Thomas. The prize itself was awarded by a close friend of Revd Thomas, Dr Revd Samuel Warneford, whose influence would only grow after this date and instigate a very different chapter in the history of the medical school.

Like the Cox family medical practice, the school expanded as the result of Edward Townsend Cox's private income. By 1838, however, this all changed when William Sands Cox induced one of his patients, whom he had treated free of charge for many years, to help fund his educational venture. Having already sponsored, if not shaped, a number of other medical institutions, including the Radcliffe Lunatic Asylum in Oxford, Samuel Warneford used his influence to enlarge the school and, in the process, transform it into what began to resemble a religious college. Nevertheless, the school was first and foremost a medical institution. Most students after all continued to enrol at the school in order to study medical subjects for varying periods. Tuition costs, approaching £30, entitled students to a full course of lectures in the winter session.[40] These included seventy lectures on anatomy with Cox and as many lectures in physiology with James Johnstone, son of Edward. Together with a hundred anatomical demonstrations taught by Cox with the assistance of James Harmer, these subjects formed what staff described as the core of medical instruction. Nevertheless, students attended as many lectures with Knowles and Johnstone on materia medica, Woolrich on chemistry and John Eccles on the subject of physic. In contrast, Ingleby still only delivered sixty lectures on midwifery, ten fewer than were provided on the subject of practical surgery, which Cox also delivered in association with the ophthalmic surgeon, Richard Middlemore. This fact alone suggests students were making more use of the Eye Hospital's cases, though they paid only for attendance at the General Hospital (£12 12s) and General Dispensary (£6 6s). Only Knowles and Davis continued to teach medical subjects in the summer, students paying 3 guineas to attend their respective series of sixty lectures, and, in Knowles's case, much time on rural excursions 'herborising' and collecting botanical specimens.[41] However, those students so desiring could also attend the Revd William Lawson's lectures on mathematics for a similar fee.

---

[39] UBSC, MS, Minute Book, 1831–8.

[40] Ibid.

[41] UBSC, Vaughan Thomas Collection, MSS 281/i/173.

Thereafter, more substantial extensions to the school and its curriculum were instigated, all of which bore the mark of Warneford's influence. A Department of Theology was originally set up to satisfy what the school's governors regarded as 'the great need of additional clergymen, particularly in such localities as that in which the college is situated'.[42] Styled Queen's College in 1843 and modelled on an Oxford college, much like Birmingham's King Edward VI School, additional buildings were added to the site at Paradise Street at a cost of £2,600, largely defrayed by Warneford. In October, the college opened students' chambers and a dining hall. Naturally, a chapel was also hastily erected. Despite earlier claims, only now did the school truly appear to offer 'parental, moral, and religious superintendence'.[43] Just as the school assumed the appearance of a college, its students also finally resembled scholars. Only days after the 1843 academic year had commenced, ten students requested the college authorities that they be permitted to wear academic dress, a request the governors were only too happy to grant.[44]

As one might have expected, the overtly Anglican institution proved unpopular in a town which had become the home of several influential dissenting families.[45] Nonconformist places of worship also outnumbered Anglican churches in these decades. In 1851, for example, Birmingham already supported fifty-four Nonconformist chapels compared to twenty-five Anglican churches.[46] Appointments to the school also appeared to favour Anglican candidates. Neither did it help that the school continued to be administered in a very private manner, as in the days when it was run out of the Cox family practice. In contrast to other local medical institutions, especially the town's three existing voluntary hospitals, the finances of the school were also often mixed with those of its backers. Furthermore, unlike the hospitals, the school grew very quickly and with little apparent caution.

BIRMINGHAM'S FIRST TEACHING HOSPITAL: THE QUEEN'S HOSPITAL

Throughout the second half of the eighteenth century, medical students had regularly walked the wards of provincial hospitals in order to supplement the lessons learned during their very practical periods of apprenticeship.

---

42 UBSC, Vaughan Thomas Collection, MSS 281/i/168.

43 Ibid.

44 UBSC, Vaughan Thomas Collection, MSS 281/i/6.

45 D. Smith, *Conflict and Compromise: Class Formation in English Society, 1830–1914* (London, 1982), 157.

46 V. Skipp, *The Making of Victorian Birmingham* (Birmingham, 1996), 116.

If not actually attending local hospitals where their masters held posts, most pupils continued to travel to London for a time in order to obtain additional clinical training and take their qualification examinations. Like that of its students, the success of a medical school in Birmingham also depended on access to clinical cases and financial support. Warneford provided the latter in 1838, albeit with strings attached. Clinical cases, on the other hand, were originally made available through access to the work of the General Hospital, an arrangement negotiated by Edward Johnstone, the medical school's first President. While this evidently satisfied the needs of the school and the examining bodies in London, the arrangement appears to have broken down with William Sands Cox's relationship with his contemporaries.

Though on good terms with an earlier generation of staff at the General Hospital, the founder of the medical school was unable to cultivate equally strong relations with their successors, most notably the town's other young surgeons. As one might expect, this would have severe repercussions on the success of his school, especially when this brought Cox into conflict with more reputable medical practitioners, such as Joseph Hodgson, who he never even attempted to win over. Holding a less prestigious post at the local Poor Law infirmary, Cox may have been influenced by jealousy when he claimed Joseph Hodgson's lectures were not clinical enough to merit him a teaching post at his school, despite the surgeon's access to cases at the General Hospital and important contributions to the understanding of aneurism.[47] Hodgson, in response to this and, no doubt, many other less public confrontations, restricted the access of Cox's students to the hospital's wards, describing their chief instructor as a 'vile specimen of humanity'.[48] As a result, in an effort to ensure access to clinical cases, Cox, using his remaining connections, initiated a campaign to build his own hospital. With allies like Warneford, he could not fail. Only two years later, in 1841, the Queen's Hospital was opened in Bath Row, located, like all previous institutions, near the centre of Birmingham. Said to be the first purpose-built teaching hospital in England, the Queen's was, with greater certainty, the first hospital in Birmingham dedicated to medical instruction.[49] It was also very different from the town's other hospitals, with the

---

[47] See, for example, J. Hodgson, *A Treatise on the Diseases of Arteries and Veins* (London, 1815).

[48] UBSC, Vaughan Thomas Collection, MSS 281/i; see pamphlet by [A lover of truth and charity], 'Some Farewell Remarks, Concerning a Clinical Hospital, dedicated, without permission, to Mr. William Sands Cox, and his scientific and humane supporter', 1839.

[49] Challengers to this claim include the Charing Cross Hospital (1823) in London, as well as Addenbrooke's (1766) in Cambridge.

exception of the General, in that it did not occupy a building originally constructed for non-medical purposes.

Although its domestic design made it appear very similar to other institutions in Birmingham, the Queen's Hospital also stood out given the amount of fanfare that accompanied its inauguration. As is to be expected, given the rivalry that existed between Hodgson and Cox, much of this was intended to dispel suggestions that Birmingham did not need another general hospital, as was suggested by the former. While generating much attention and publicity material, the celebrations are not captured in the pages of the hospital's extant ledgers, though a comprehensive description survives in an address that was delivered by Vaughan Thomas on the inaugural day and printed subsequently.[50] On 18 June 1840, the day the hospital's foundation stone was laid, 300 members of Provincial Grand Lodge of Warwickshire and various freemason lodges representing other counties gathered in full costume for breakfast in the committee room of the Town Hall. The event was presided over by Earl Howe, the Deputy Grand Master of Warwickshire Provincial Grand Lodge of Free and Accepted Masons. In total, the breakfast was attended by 450 individuals and enlivened by a musical performance by the celebrated Distin Family, one of England's first brass bands.[51] The novelty of this event and so many distinguished guests, not to mention the performance of the country's premier saxhorn quintet, led many ordinary people from surrounding villages and hamlets to visit Birmingham on the day.

After breakfast, official guests removed to Bath Row by way of a circuitous route. The assembled party then commenced to parade through the town's streets ending their procession at the hospital grounds and, despite some obvious fears of disorder, 'not the slightest confusion or accident occurred to mar the auspicious commencement of this noble and benevolent work'. On the right hand side of the enclosure, at the corner of Bath Row and Falconer Road, a large temporary platform had been erected immediately above the spot where the foundation stone was laid and blessed by Revd Dr Marsh to accommodate a group of invited spectators, including members from principal families in the neighbourhood, as well as hospital subscribers. The majority of the guests occupied the ground in which the future building would stand. Conducting a Masonic ritual that would have appeared very familiar to those who had attended similar ceremonies, Earl Howe poured corn, wine and oil on the stone, these items representing to his audience 'all the necessaries, conveniences, and comforts

---

[50] V. Thomas, *An Address upon Laying the Foundation-Stone of the Queen's Hospital, Birmingham, June 18, 1840* (Oxford, 1840).

[51] M. Mamminga, 'British brass bands', *Music Educators Journal* 58, 3 (1971), 82.

of life'. At this point, Vaughan Thomas delivered his address, beginning
with the suggestion that the foundation stone of the hospital had been 'so
geometrically squared, and so skilfully laid, that it may well serve for an
emblem of the fortunes of our future institution, of its stability as a Temple
of Learning as well as Mercy' and 'of the rectitude of its administration'.
He subsequently recounted a feminine tradition of charity extending from
the fourth-century Roman saint, Fabiola, who is to have built the first hospital in the West, to Queen Victoria, the hospital's patron. Thomas then
drew attention to more recent changes in medical education, describing the
institution as one of the country's first teaching hospitals. Given the novelty of such a notion, he proceeded to comfort his audience by declaring
that the 'educational [function of the hospital would] be kept in due sub-
ordination to its Charitable purposes', stating that science was 'the hand-
maid, not the rival or the ruler, of Charity'. On concluding his few words,
the band of the Royal Scotch Greys played 'God save the Queen', and the
Masons ceremoniously returned to the Town Hall, where the Provincial
Lodge was formally closed.

The hospital first opened to patients a year later in 1841 with seventy
beds. During the next twelve months, staff treated more than 500 inpatients and three times as many outpatients.[52] Over the next five years, inpatient numbers doubled and outpatients surpassed 2,200.[53] Unfortunately,
few other details concerning the work of the institution during this period
exist, as early hospital records are exceedingly scarce prior to 1849, when
governors issued their first printed report. The most significant developments, however, are recounted in late nineteenth-century annual reports,
which, as at other hospitals, often recount the most significant dates in a
charity's history. In the case of the Queen's Hospital, the next such significant milestone was attained in 1845, when a seventy-bed fever ward
funded by several Staffordshire ironmasters was added to the building.[54] As one might expect in an age preceding germ theory, the block
was located some 300 feet behind the main building, lying to the north
of the hospital, nestled between the patients' recreation ground and the
Worcester & Birmingham canal. While the ward enhanced the charity's
services to the community considerably, the hospital's central location
and proximity to wharves and railway depots also ensured it attracted a
disproportionate number of accident cases. Despite expectations, medical staff claimed the majority of fractured jaws, legs and ribs occurred
as a result of either drunken fits and fighting, as opposed to working

[52] BCLLS, Queen's Hospital, Annual Report, 1848.
[53] Ibid.
[54] BCLLS, Queen's Hospital, Annual Report, 1852.

conditions.[55] Similar opinions are expressed in the clinical notes compiled
by students who worked under the supervision of the hospital's six medical
officers.[56] Besides Sands Cox, the institution's two other surgeons in these
years were George Knowles and Langston Parker, an expert on syphilis. Its
three physicians included Samuel Wright, David Nelson and Birt Davies,
who also served as the borough's first coroner. As at other hospitals already
described, the majority of the day-to-day work was undertaken by a dispenser, a matron and their staff, in this case comprising nine nurses, a house
maid, a cook, two kitchen maids and three porters. Though resembling the
town's other medical institutions, the hospital, like the medical school with
which it was associated, was at times run like a private business, many of
its rules and regulations being flouted by its primary consulting surgeon,
William Sands Cox.

At first, this hardly appeared to hinder its development. To cope with
increasing patient numbers, the hospital surgery and dispensary were
enlarged in 1850. The building also gained an additional operating theatre
and two wards for operative cases, chloroform being used in all operations;
the first administration of the anaesthetic was performed by Langston
Parker on twenty-two-year old Mary Ann Chambers, who had her foot
amputated by George Knowles.[57] Included in the building's alterations
was a committee room, chapel and house steward's room, the expense of
construction having been defrayed by an exhibition of 'specimens of high
art and manufacture', as well as musical performances and, more unusually, an Artisan's Fund, to which many local workshops contributed from
at least the mid-1840s.[58] Since the hospital had opened, 230 students had
walked its wards, and approximately 40,000 patients had been 'cured or
relieved'.[59] During the same period, the number of beds at the institution
had more than doubled, reaching 150, approximately 100 less than were in
use at the General Hospital.

With its own hospital and a very generous source of private funding,
Queen's College, to many observers, especially those residing outside
Birmingham, appeared to be the leading provincial medical school. Each
year, approximately a dozen of its students were obtaining the diplomas
of the Royal College of Surgeons, while an equivalent number successfully passed the examinations of the Society of Apothecaries. From £186

[55] Ibid.

[56] BCLLS, J. Moore, 'Clinical Reports of Surgical Cases under the Treatment of
     W. Sands Cox, 1843–4' (Birmingham, 1844).

[57] *Birmingham Journal*, 16 January 1847.

[58] BCLLS, Queen's Hospital, Annual Report, 1853–4.

[59] Ibid.

in 1841, income from student fees rose annually, peaking at £550 in 1851.[60] In 1846, preparatory education had commenced at Queen's College, in order to fulfil the conditions of existing corporate and collegiate educational bodies. With the appointment of Oxbridge-educated tutors in classics and mathematics, as well as an instructor in charge of the college's junior department, the school was permitted to issue the degrees conferred by the University of London on the same footing as the newly founded King's and University colleges, London. Then, just as Birmingham's college seemed poised for greater success, student numbers dropped abruptly, tuition fees in 1853 raising just £133.[61] By 1860, income declined still further to an embarrassing £13. By subsequently borrowing money to which he was not entitled and exercising excessive power over the appointment of officers regarded to be sympathetic to the college, Cox caused the hospital further inconvenience, as did a subsequent decline in the number of hospital subscribers.

Having sensed a potential niche in the market, not to mention a certain amount of dissatisfaction with the school's administration, staff at the General Hospital established a rival school. Named Sydenham College, after the seventeenth-century physician variously described as the father of English medicine and the English Hippocrates, the school opened its doors in 1851. Though occupying less imposing buildings in St Paul's Square, its medical curriculum was identical to that offered at Queen's College, with the exception of a clinical course on insanity taught by George Bodington of Sutton Coldfield, who kept a private asylum. With a student body said to be 'attentive' and 'eager to imbibe the information imparted to them', and lecturers 'taking pains to make their subjects understood by the students', Birmingham's second medical school began to attract greater numbers of students at a time when Queen's College's difficulties only appeared to commence.[62] By 1854, Sydenham staff began their own series of inaugural lectures to promote the college and attract potential provincial medical men to Birmingham, a feat the school appeared to be achieving. Though surviving documents relating to the school provide no details concerning exact student numbers, a report in 1857 lists a dozen prize-winning students alone, and the following year, the school moved from St Paul's Square to larger, more suitable buildings opposite the General Hospital.[63] With the appointment of a medical tutor in 1860 to assist first-year medical students, numbers were again set to increase with patient numbers, which surpassed

[60] BCLLS, Queen's Hospital, Annual Report, 1852.

[61] BCLLS, Queen's Hospital Annual Reports, 1848–60.

[62] UBSC, Minutes of Sydenham College, Birmingham, 1851–66, MS 128/1.

[63] Ibid.

23,000.[64] Though short on detail, a further report in 1861 claimed students entering the institution that winter were double the usual number.[65] From this point, any reverse in the steady decline of student numbers at Queen's College appeared unlikely.

Not all of the school's decline, however, can be attributed to the existence of a local rival. Clearly, much of the school's misfortune appears to have been the result of unforeseen transformations at Queen's College, where clerical education had superseded clinical instruction. In contrast to Grainger's school of anatomy in London, which had commenced life at St Saviour's Churchyard before its move to Webb Street, Cox's school appeared to develop in reverse order. A once fledgling medical school had been saved by a generous patron, but was now linked to a rapidly expanding theological college. Though a medical tutor had been appointed in 1846 to teach the various branches of medicine, surgery and midwifery, he was required to take his meals with students and attend regular services in the chapel.[66] A year later, the college also began to grant certificates in the arts and law. The duties of the arts warden, however, suggest this was simply another branch of religious instruction. Besides being required to examine students for entry to all departments, the warden was responsible for the instruction of every student in Christian ethics, church history and the doctrines of the Church of England.[67] All students attended chapel, where prayers were read daily and divine service was performed twice every Sunday, with two sermons, both of which pupils were to attend. Finally, in 1850, after further additions to the site, a Professor of Pastoral Theology was appointed and a Department of Theology founded. Presumably, this was also the extent of Warneford's original vision, for he gave the college a final gift of £12,000 in Great Western preference stock in order to guarantee the future salaries of the most recently appointed instructors.[68] Although the trustees had other plans, most assumed that by indulging this wealthy benefactor only greater sums would be forthcoming at his death. In fact, Warneford lived only three more years, and, what must have been the council's greatest shock, he left them with nothing.

During his final years, Warneford decided that the spirit of his intentions in Birmingham, despite his best efforts, would be 'perverted by

---

[64] BCLA, GHB 417.

[65] UBSC, Minutes of Sydenham College, Birmingham, 1851–66, MS 128/1.

[66] BCLA, An Act for the Regulation of Queen's College at Birmingham, 12 April 1867, 157.

[67] Ibid., 158.

[68] Ibid., 166.

posterity'.[69] Regular conflict with staff only served to convince him that others would be less vigilant than he and more easily tempted to drop their guard against what he referred to as 'future satanic subtlety'.[70] By leaving the college with nothing, however, he merely ensured his prophesy, as he had helped develop and expand the school's facilities without securing future funding. Denied such healthy prospects, Queen's College began a period of decline and, within a decade of Warneford's death, found itself on the 'Verge of Bankruptcy'.[71] By 1864, the relationship between Queen's College and its teaching hospital was finally dissolved following a thorough investigation by the Charity Commissioners, who condemned the secretive way in which both school and hospital were run.[72] As a result, the college was required to relinquish trusteeship of the hospital's funds, and its role in the appointment of chaplains, medical and surgical officers as well as the distribution of fees for clinical instruction. Moreover, in 1868, following additional investigations and negotiations, the town's two medical schools merged, signalling the start of a new, dynamic period for medical education in Birmingham.

The merger's success was apparent from the outset. Records were finally properly maintained, and both institutions were run according to rules and regulations rather than the whims of founders and key donors. As a result of this and a more inclusive policy generally, besides regaining the confidence of Birmingham's dissenting community, the school became more popular with both the public and potential medical students. The hospital, in turn, remained a favourite of the town's workers as it continued to treat many work-related accidents. In appreciation of this service, an Artisan's Fund had been originally set up in 1846 by a Mr S. Bradley, its members regularly meeting at the Hope and Anchor Tavern on Navigation Street. The following year, the fund's managers donated nearly £1,000 to the institution.[73] During the 1870s, with the help of certain influential inhabitants, including one of the hospital's surgeons, Sampson Gamgee, and its new Secretary, Henry Burdett, the scheme was developed into the Birmingham Hospital Saturday Fund. Initially paying for an additional outpatients' department, a hospital laundry and a small pathological block at the Queen's Hospital in 1873, the scheme financed many other important developments at the institution, as well as other local

[69] B. Windle and W. Hillhouse, *The Birmingham School of Medicine* (Birmingham, 1890), 3.

[70] Ibid.

[71] BCLLS, Queen's College, Birmingham, Act, 1867, 169.

[72] BCLLS, Queen's College, Official Report to the Charity Commissioners, 1863.

[73] BCLLS, Queen's Hospital, Annual Report, 1853–4.

hospitals.[74] By the 1880s, regular workplace collections had emerged in some forty English provincial centres and would begin to provide ever greater percentages of voluntary hospitals' ordinary income.[75]

Though the importance of the Hospital Saturday Fund to local medical institutions was not yet fully recognised, other changes due to the institution's reorganisation were more palatable. The laying of the foundation stone at the Queen's Hospital's new block on 4 December 1871, for example, is worth describing in greater detail, given the way in which it highlights the broad support the newly inaugurated charity enjoyed within the local community. In addition, the 1871 event contrasts noticeably with the celebrations held at the institution in 1840 and described by the Revd Vaughan Thomas. While the earlier ceremony was an opportunity for Birmingham's worthiest and wealthiest families to associate themselves with a traditional form of paternalist welfare, the latter ceremony was clearly intended to underscore the benefits of working-class self-help. Furthermore, in an effort to ensure the attendance of substantial numbers of workers, Robert Baker, the town factory inspector, agreed to make Saturday, 2 December an ordinary day of work, rather than a half day of work, in order to make the following Monday, 4 December, a half holiday.[76]

On this occasion, instead of being led by freemasons, the procession to the hospital site began with the police, their marshals and a brass band, who were immediately followed by local postal carriers, the Gem Street Industrial School Brass Band, the General Committee of Working Men's Extension Fund, before a number of firemen in full uniform. This first group was followed by the Mitre Cut Nail Works Brass Band, a display of banners, members of the Independent Order of Odd Fellows (Manchester Unity), more bands and banners and the Birmingham District Branch of the Ancient Order of Foresters. Members of these and other friendly societies were followed by yet more bands, banners and friendly societies, including the Independent Order of Good Templars, the Universe Works Brass Band, Rope and Twine Spinners, the Grand United Order of Odd-Fellows (Birmingham District), as well as the Penn Street Industrial School Drum and Fife Band. Finally, the procession closed with a two-horse carriage on which a printing press was drawn. The same press had been used to print an image of the new buildings which a group of volunteers distributed along

---

[74] BCLLS, Queen's Hospital, Annual Report, 1873 and 1874.

[75] S. Cherry, 'Hospital Saturday, workplace collections and issues in late nineteenth-century hospital funding', *Medical History* 44, 4 (2000), 486–7; M. Gorsky and J. Mohan with T. Willis, *Mutualism and Health Care: British Hospital Contributory Schemes in the Twentieth Century* (Manchester: Manchester University Press, 2006), 18–42.

[76] BCLA, HC/QU/1/1/1.

a parade route that started at St Martin's vegetable market, a public square. From this central location, the procession journeyed by way of Jamaica Row to the Bull Ring, travelled down New Street to Paradise Street, followed East Row and Broad Street, then proceeded through Islington and Islington Row and finally marched down Bath Row to the hospital ground at approximately 2:30 pm, the whole journey having taken an hour. Unlike the ceremony staged in 1840, parade organisers on this occasion were disappointed that the hospital gate was not opened for 23 minutes after the procession had begun to arrive at the hospital grounds. This regrettable oversight led to a general break-up of the marchers and great injury to a number taking part, the head group having been at the gates for between three and four minutes 'before the crush began'.[77] Besides this unfortunate episode, which immediately tested the skills of the newly appointed hospital staff, the procession was judged a success and, in contrast with events staged in a previous era, cost the organisers only £8. Much more was no doubt spent on a soiree held afterwards and attended by 1,100 guests at the Town Hall, where one of the many bands in attendance on the day performed for their unusually large audience. A subsequent glee party was followed by a final inspiring address by the Nonconformist minister George Dawson, regarded by historians as the leader of the Civic Gospel movement and 'the greatest talker in England' by his contemporaries.[78] The day concluded with much dancing.

As this should suggest, by the 1870s, the town's medical charities appeared more than ever to belong to the people of Birmingham. Not surprisingly, by 1875, the finances of the Queen's Hospital had improved to an extent that allowed its governors to make the hospital a free institution, whereby patients no longer required either letters of recommendation or tickets from subscribers in order to access medical care.[79] While generously supported and justifiably admired by local residents, it also attracted some attention outside Birmingham. For example, Benjamin Ward Richardson, the physician and self-proclaimed disciple of Chadwick considered the charity to be a model hospital. In *Hygeia, A City of Health* (1876), his treatise on an imaginary garden-city of health, originally delivered in 1875 as an address to the Social Science Association in Brighton, Richardson refers to the institution when considering the ideal hospital:

> The out-patient department, which is apart from the body of the hospital, resembles that of the Queen's Hospital, Birmingham, – the first out-patient department, as far as I am aware, that ever deserved to be

77 BCLA, HC/QU/1/1/1
78 Hopkins, *Birmingham: The Making of the Second City*, 52.
79 BCLA, HC/QU/1/1/2.

seen by a generous public. The patients waiting for advice are seated in a large hall, warmed at all seasons to a proper heat, lighted from the top through a glass roof, and perfectly ventilated. The infectious cases are separated carefully from the rest. The consulting rooms of the medical staff are comfortably fitted, the dispensary is thoroughly officered, and the order that prevails is so effective that a sick person, who is punctual to time, has never to wait.[80]

At the real hospital that same year, the sale of investments, totalling £10,000, was used to pay for the institution's further enlargement. Still equipped with only 130 beds, the hospital treated 1,828 inpatients, but offered a variety of medical services to more than 14,000 outpatients, allowing it very easily to fulfil its important teaching functions. Quite appropriately, the same annual report that announced this very new direction in the delivery of hospital services also conveyed the death of Sands Cox.[81] As a result, the events gave the impression that a new era in Birmingham medicine was about to commence.

[80] B. W. Richardson, 'Hygeia, a city of health', *Nature* (12 October 1876), 542–5. Thanks also to Graham Mooney for drawing my attention to this fascinating document, and Clare Hickman for her insight into Richardson.

[81] BCLLS, Queen's Hospital, Annual Report, 1875.

CHAPTER 4

Mid-Victorian Specialties:
Hospitals for Women, Children and the Deaf

THE 1830s have been described as an usually quiet period for hospital construction, as the moneyed classes throughout England sought alternative ways to manage the requirements of the sick poor. In general, a reformed Poor Law was seen to meet the needs of the destitute in large industrial communities more effectively than hospitals and dispensaries which often appeared to allocate their services indiscriminately.[1] The decades following the New Poor Law of 1834, however, were characterised by an intensified period of hospital foundation, especially specialist institutions. Though opponents to specialisation remained vocal throughout the 1860s, members of the medical profession had begun to recognise specialties as proper and legitimate fields of practice. A decade later and specialisation was already regarded as inevitable.[2] As a result, during this second wave of voluntary hospital foundation many of the country's largest cities and towns gained specialist hospitals, such as those for women and children, while other towns, which had long survived without institutions for those with specific diseases or disabilities began to develop their specialist services. For example, between 1834 and 1861, fifteen more eye institutions are known to have been established throughout England, and eighteen more were founded in the next two decades.[3] In London, twenty-two specialist institutions were founded in the 1860s alone.[4] By 1861, more than a hundred specialist hospitals existed nationally.[5] Between 1843 and 1871, Birmingham gained several, including hospitals for the deaf, children and women. Not all thrived, but like those located in the capital and other industrial towns, these institutions rapidly attracted patients, many more coming from great distances given improvements in transport. Unlike earlier institutions, most opened with beds for inpatients and grew more quickly than did a first generation of English specialist hospitals.

---

[1] Pickstone, *Medicine and Industrial Society*, 84.

[2] Rosen, 'Changing attitudes', 352.

[3] Sorsby, 'Nineteenth century provincial eye hospitals', 507.

[4] M. J. Peterson, *The Medical Profession in Mid-Victorian London* (Berkeley, 1978), 262–3.

[5] R. Pinker, *English Hospital Statistics, 1861–1938* (London, 1966), 57, 61–2.

A SMALL, BUT SOUND MEDICAL CHARITY:
THE EAR HOSPITAL

By the second half of the nineteenth century, charitable initiatives in Birmingham were supporting a greater number of local medical charities. For example, only three years after the completion of the Queen's Hospital, another very different, specialised institution was inaugurated. The Ear and Throat Infirmary in Birmingham was founded in 1844 largely through the efforts of Mr William Dufton, almost three decades after the first English institution devoted exclusively to the treatment of diseases of the ear opened in London's Soho district in 1816. Earlier provincial ear hospitals appeared in Shrewsbury (1818), Liverpool (1820) and Leeds (1821), but most of these institutions had wider remits than their names suggested, and treated diseases of the eye under the same roofs.[6]

Before the existence of aural institutions, the medical profession seemed content to carry out only superficial examinations of the external ear, leading many people who suffered from afflictions of the inner ear, like those with ailing eyes, to look beyond orthodox medicine for treatment. According to the founders of ear hospitals, the remedies generally prescribed by the trained practitioner seldom gave any relief, and the public, not surprisingly, regularly appealed to empirics in cases of deafness. No one could blame members of the medical profession for not wishing to tamper with so important an organ as the ear, situated as it is so close to the brain. In all likelihood, many who had endeavoured actively to intervene with diseases of the ear may just as often have brought about permanent deafness, if not death, in their patients. As a result, for the same reasons one would not commit a watch requiring repair to the hands of any but the watchmaker, Dufton discouraged the public from attending those who were ignorant of the ear's anatomy.[7] In this way, the inhabitants of Birmingham were encouraged to support one more voluntary hospital and, over the course of the next five decades, many other specialty institutions would follow, some more popular than others.

By this time, few members of the public were entirely ignorant of ear ailments, and especially deafness. For example, many of Birmingham's inhabitants had been introduced to the science of instructing deaf and dumb persons by Dr Gabriel DeLys, whose involvement in the Eye Hospital has already been alluded to. In a lecture delivered at the town's Philosophic

---

[6] R. Kershaw, 'British ear and throat clinics historically considered', *Journal of Laryngology, Rhinology and Otology* (1913), 423.

[7] W. Dufton, *The Nature and Treatment of Deafness and Diseases of the Ear* (Birmingham, 1844), xii.

Institution in 1810 on the subject of hearing, the French émigré and member of the local Humane Society illustrated his lesson by introducing the audience to one of his pupils, Jane Williams, an eight-year-old girl, who had been deaf and dumb since birth. Clearly impressed with what appeared to be a 'very engaging and intelligent child', not to mention the importance of the subject, members of that particular audience subsequently formed a society for the education of similarly affected children between the ages of eight and thirteen.[8] Two years later, a suitable building with accommodation for forty children of both sexes, with playgrounds and a master's residence was established in the parish of Edgbaston on grounds donated by Lord Calthorpe. There, children learned skills that would allow them to earn a livelihood and, as importantly, no longer live 'in the utter ignorance of God'.[9] By 1858, the institution underwent some needful alterations allowing it to receive 120 pupils, many from the Midlands, especially West Bromwich and Northampton, where associate branches were formed, but often from much greater distances.

Unlike the town's deaf and dumb institution and much like earlier specialist institutions, Birmingham's Ear Infirmary grew less rapidly after its foundation in 1844. Founded six years after James Yearsley founded the country's first ear, nose and throat hospital in Sackville Street, Piccadilly, the Birmingham ear dispensary moved frequently in subsequent years, operating out of premises at Cherry Street in 1863.[10] At this time, the charity was treating 2,077 patients annually, a third of whom came from outside Birmingham, at a cost of only £120, approximately a fifth of which was spent on medicines and hearing instruments.[11] Besides announcing the appointment of John Adams to the post of dispenser, reports in these years include some statistics relating to the charity's first patients. For example, of approximately 376 deaf patients, nearly half had lived with their affliction for five years or more. Like many specialist hospitals, a small staff, comprising perhaps a single surgeon or physician, devoted many years initially treating large numbers of chronic cases. Even more remarkably, of the 629 patients discharged in 1863, more than 300 were said to have been cured, only 100 being regarded as incurable. Together, these figures clearly suggested both the need for such an institution locally, as well as the staff's proven ability to treat patients effectively.

Though rarely described in annual reports, the actual treatment of those

[8] G. Griffiths, *History of the Free-Schools, Colleges, Hospitals* (London, 1861), 116–17.

[9] Ibid.

[10] R. S. Stevenson and D. Guthrie, *A History of Oto-laryngology* (Edinburgh, 1949), 63.

[11] BCLLS, Birmingham Ear Infirmary, Annual Report, 1863.

attending the dispensary – at least those attending for aural complaints – is captured in the pages of a publication by the hospital's first surgeon. William Dufton's *The Nature and Treatment of Deafness and Diseases of the Ear* (1844), published by the aural surgeon in the same year he helped establish the Ear Infirmary, was described by the *Lancet* in 1845 as 'one of the best compendiums of aural medicine and surgery which has hitherto been published in this language'.[12] As a historical source, its value is twofold: it discusses a number of the charity's earliest cases and clearly conveys the prevailing ideas of disease causation and treatment in the decades preceding the emergence of germ theory in the field of otolaryngology, not to mention a greater understanding of the workings of the ear. In general, Dufton divided the principal diseases of the ear into three types. These included the chronic and acute inflammation of the general ear, secondly, the inflammation of its particular parts, including the growth of tumours and the introduction of foreign bodies into the ear, and, finally, those diseases not being inflammatory, which he regarded as a reflection of the peculiar state of aural nerves. Of the first mentioned ailments, otitus, or inflammation of the ear, was described as one of the most common diseases to afflict his patients.[13] In most cases, treatment involved the administration of saline purgatives and an antiphlogistic diet, as was commonly prescribed for other patients suffering with various forms of inflammation. The most serious cases of inflammation may have induced Dufton to shave patients' heads, puncture the skin of swollen ears, apply blisters or leeches to the forehead and behind the ears, poultices to the affected areas, as well as regularly syringing the aural canal with warm water. More invasive approaches and treatments remained rare until the last decade of the nineteenth century. Surgical cures in these years, for example, tended to be limited to the excision of polyps and the puncture of the tympanic membrane. The latter operation in particular was advocated as a cure for deafness in the early 1800s and carried much weight in following decades despite much opposition because it was championed by Sir Astley Cooper, one of the most influential surgeons of the day.[14] Far more often, Dufton's patients endured regular irrigation and repeat examinations, including the insertion of probes as he monitored the progress of their cases, as well as the effectiveness of his remedies. Alternatively, catheters were often employed to inject water, air or vapours, including tobacco smoke, into the middle ear. Should his methods have proved unsuccessful and a patient's life lost, his case notes suggest post-mortems were occasionally undertaken,

[12] *Lancet*, 13 September 1845.

[13] Dufton, *Nature and Treatment of Deafness*, 19.

[14] Stevenson and Guthrie, *History of Oto-laryngology*, 46.

most uncovering massive infections often extending to the inner regions of the brain.

Approximately a dozen cases in his treatise on the ear also suggest his patients were very young, if not very old. Many common childhood fevers, for example, including scarlet fever, measles, smallpox and other afflictions of the skin commonly affect the ear. Neither was the introduction of foreign bodies into ears an infrequent occurrence, especially amongst children at play. Dufton claimed such obstructions and diseases among infants were easily detected with their help, as the children on examination habitually grasped their affected ears.[15] Interestingly, like Darwall and other local practitioners writing in these years, he is somewhat sceptical about patients' accounts of illness. Primarily, he suggests the statements of patients and friends were rarely to be relied on, claiming many parents insisted children had thrust various objects into their ears, only to find nothing upon closer investigation.[16] The importance of this advice is then highlighted with a cautionary tale concerning a child who was forcibly restrained while medical staff at an unnamed public institution fished for foreign objects in his ear using a pair of tooth-forceps finding only the boy's malleus, or hammer – the largest of the three ossicles of the middle ear – which his operators inadvertently removed. The child is said to have died a few days later from a brain inflammation.

Despite such warnings, Dufton's treatise reveals that many of the children he treated suffered over extended periods. For example, many of his youngest patients endured their afflictions for months, while a substantial number reported aural suppuration over several years. In many cases, such long periods of suffering would have been encouraged by a belief prevalent among a previous generation of surgeons that purulent discharge from the ear, or any wound for that matter, should be encouraged rather than suppressed. This tradition was eventually challenged in the first decade of the nineteenth century by Moorfields founder John Cunningham Saunders, then a demonstrator in anatomy at St Thomas's Hospital, London, who, in his published work, *The Anatomy of the Ear* (1806), advised surgeons to syringe purulent ears with a solution of zinc sulphate.[17] Given frequent references to deafness caused by the presence of hardened masses of wax and various foreign substances, such as wool, dust and even insects, it comes as no surprise that simple cleanliness, as argued by another ear specialist, in

---

[15] Dufton, *Nature and Treatment of Deafness*, 51.

[16] Ibid., 89–90.

[17] Stevenson and Guthrie, *History of Oto-Laryngology*, 47. Three years later, Cunningham founded the institution which would become Moorfields Eye Hospital.

this case William Wilde of Dublin, the father of Oscar and founder of St
Mark's Hospital for the Eye and Ear, remained central to treatment at ear
hospitals for much of the nineteenth century. As a result, in their earliest
years, staff at these charities also posted unusually high rates of cure, especially for deafness.

Despite a noticeable expansion in the numbers of patients treated
at the institution and attracting the patronage of Lord Calthorpe, these
were difficult years for the charity. The first annual reports issued by the
charity in the 1860s indicate that a £50 legacy left by a Miss Primer was
the single largest donation to the institution between its foundation
in 1844 and Dufton's death in 1859. On his death, the founder was succeeded by Charles Warden, a surgeon who introduced the laryngoscope
to the Ear Hospital and thereby attracted greater numbers of patients suffering from throat rather than purely ear afflictions. Invented by Manuel
Garcia, a Spanish tenor and singing instructor, the laryngoscope did for
oto-laryngology what the ophthalmoscope did for ophthalmology. The
impact of this particular technology on specialisation is just as easily
demonstrated in hospital records. According to an annual report of 1863,
throat ailments now headed the list of cases treated at the institution, followed by diphtheria, ulceration of the tonsils, inflammations of the throat,
mouth and nose, polypi of the nose and ears, and only then eruptions of
the ears. Though neglected by Warden's forerunners, the diseases of the
larynx as understood before the invention of the laryngoscope had been
studied extensively by another local practitioner, Frederick Ryland, who
received the Jacksonian prize in 1835 for his efforts.[18] This example alone
should remind historians that the origins of medical specialties in towns
like Birmingham almost always pre-date the foundation of a specialist
hospital.

An increase in the charity's work in these years is further indicated
by the appointment of a dispenser in 1863, and some additional prestige
undoubtedly came to the charity with the appointment of Joseph Toynbee
as the institution's consultant surgeon. Toynbee, who sought to develop
the field of aural pathology by conducting thousands of dissections while
assistant curator under Robert Owen at the Hunterian Museum in the
1840s, became the first aural surgeon to obtain a position in an English
general hospital when a special ear department was established at St
Mary's Hospital, London in 1851.[19] While there, Toynbee undoubtedly

---

[18] F. Ryland, *A Treatise on the Diseases and Injuries of the Larynx and Trachea*
(London, 1837).

[19] E. Heaman, *St Mary's: The History of a London Teaching Hospital* (London, 2003),
45.

conducted much of the research that resulted in his magisterial study, *Diseases of the Ear* (1860), though some of the cases outlined in the text might have been encountered while on a rare visit to Birmingham. Though statistically richer than Dufton's work, Toynbee's classic text was not that different from the work produced by the Birmingham surgeon. For example, both surgeons treated otitis by applying 'leeches behind the ear' and with 'copious syringing'.[20] Additionally, both works concluded with chapters on nervous deafness, which was a particular concern in these years and variously attributed to over-study, want of sleep, gout and typhus fever.

Despite a noticeable increase in knowledge relating to diseases of the ear, nose and throat, the financial support of the Birmingham Ear Hospital appeared relatively static in the institution's first decades. In 1866, the hospital still had fewer than seventy subscribers, while staff continued to attend more than 500 patients annually.[21] The fact that each subscriber was entitled to send eight patients to the charity suggests that nearly every donor distributed their permitted limit of admissions letters. In the same year, the hospital lost the services of Toynbee, who died as the result of chloroform experiments he conducted on himself in St Mary's chemistry laboratory in an effort to cure his tinnitus.[22] Difficulties were exacerbated in 1868 by a poorly planned move to Ann Street, which also perhaps highlights the importance of advertising medical charities, something few medical practitioners formally encouraged given its association with quack showmanship. Unaware of the hospital's relocation on this occasion, those patients who continued to turn up at the charity's old premises became convinced of its closure and ceased attending altogether.[23] Presumably their numbers comprised primarily those patients who resided outside of Birmingham. Some of them may also have begun to attend a rival hospital where they were attended by Dr Alexander Fleming and Mr Benjamin Hunt; the Birmingham Eye and Ear Hospital and Free Dispensary, Birmingham's second specialist ear institution, occupied premises in Cambridge Street in the mid-1860s, before moving to Suffolk Street.[24] Consequently, in order to ensure the survival of the main aural charity, its administrators suggested a merger with the Eye Hospital, which had purchased the spacious premises formerly known as Dee's Royal Hotel in 1861. The decision

[20] Stevenson and Guthrie, *History of Oto-laryngology*, 67.

[21] BCLLS, Ear Hospital, Annual Report, 1866.

[22] Heaman, *St Mary's*, 45.

[23] BCLLS, Ear Hospital, Annual Report, 1869.

[24] *Birmingham Red Book and Reference Almanac*, 1867–71. Many thanks to Andrew Edwards for drawing my attention to this hitherto overlooked institution.

to merge appeared very logical as there had always been some overlap in the cases these two medical institutions treated. In fact, of the first dozen ear hospitals to open in the British Isles, nine were joint ear-eye institutions.

As this might suggest, the decision to operate from the same premises caused fewer difficulties than the Ear Infirmary's previous move, as staff at each charity saw their patients, few of whom were inpatients, only twice a week and could therefore contemplate sharing a building with another medical charity that treated primarily outpatients. Between 1870 and 1880, patient numbers tripled and finances improved despite a further decline in subscriptions. Largely this increase resulted from the introduction of supplementary tickets in 1879, which were valid for eight weeks (six weeks after 1881) and sold for 3s 6d a piece.[25] Nevertheless, the provincial ear hospitals at this time were less than half the size of their metropolitan counterparts, which treated approximately 5,000 patients annually.[26]

By 1881, 1,672 patients were being treated at the hospital. Unlike the town's other voluntary hospitals, the majority of patients were women. Of the 1,672 patients seen in 1881, 789 (47 per cent) were males, while 883 (53 per cent) were females.[27] A particularly detailed annual report further permits one to break down the population by parish, age and affliction. For example, the majority of 'country' patients came to the institution from West Bromwich, which sent thirty-five (4.7 per cent) individuals, the next greatest number coming from Kidderminster (eighteen, or 2.3 per cent). Interestingly, the age distribution of patients did not mirror that of other hospitals in the town either, given that the charity treated more than the average number of individuals over the age of fifty. In general, the town's eldest inhabitants up to this date appear to have been neglected by the voluntary hospitals and were disproportionately well represented in the populations of Poor Law institutions, where they usually comprised approximately a third of inmates.[28] They also tended frequently to suffer from deafness. Nevertheless, younger patients were also unusually well served by the Ear Hospital. While only ten of the ear charity's patients were less than a year old, more than 500 (31 per cent) were under the age of fifteen.[29] As at most other hospitals, the bulk of the charity's cases were working-age adults, whose numbers comprised more than a thousand

[25] BCLLS, Ear Hospital, Annual Report, 1879.

[26] G. Gould, *A History of the Royal National Throat, Nose and Ear Hospital, 1874–1982* (Ashford, 1998), 13.

[27] BCLLS, Ear Hospital, Annual Report, 1881.

[28] BCLA, Birmingham Board of Guardians, Minute Book, 1866, GP/B/2/1/33.

[29] BCLLS, Ear Hospital, Annual Report, 1881.

individuals between the ages of sixteen and fifty, leaving just 150 (9 per cent) over the age of fifty, a small but significant group.[30] Patients' particular ailments again justified the charity's claim to be an ear, nose and throat infirmary. Primarily, staff treated 779 diseases of the external ear, 538 of the middle ear, and 619 that affected the throat and middle ear. Remaining cases included patients suffering from diseases of the nose (71), mouth (28), inner ear (94), throat (31), neck (88) and jaw (14), leaving 129 miscellaneous emergencies.[31]

The institution's work continued much the same until 1883 when changes at the Eye Hospital instigated another move. Clearly evolving more rapidly than the aural side of the charity, patients and subscriptions respectively necessitated and permitted the governors of the Eye Hospital to construct a new hospital at a time when the Ear Infirmary still had only 2,000 cases on its books. Though founded within two decades of each other, the Eye Hospital had become the more successful institution, while the ear charity reverted to premises formerly occupied by the Orthopaedic Hospital before moving to 7 Great Charles Street in 1884. The Eye Hospital's spectacular rate of growth came about largely because its cases were seen as less chronic than those treated by the town's auditory specialists, though the Ear Hospital's annual reports confidently claimed a cure rate that averaged 75 per cent throughout these years.[32] As a result, the hospital's plight can again be seen merely to highlight the primacy of sight over all other senses among members of the Victorian public.

CARING FOR THE INNOCENT: THE CHILDREN'S HOSPITAL

While many of the town's inhabitants may have questioned the need for an ear infirmary, a hospital for children was more easily justified. For example, their numbers in society at the time alone appeared to warrant the establishment of such an institution, more than a third of the population of Victorian England having been under the age of fourteen.[33] The foundation of such an institution also followed a more general pattern of

[30] The report gives their exact numbers as: 10 under the age of a year, 136 under 5 years, 187 between the ages of 6 and 10, 184 between 11 and 15, 259 between 16 and 20, 360 between 21 and 30, 236 between 31 and 40, 149 between 41 and 50, 90 between 51 and 60, 48 between 61 and 70, and the remaining 13 between 71 and 80 years of age.

[31] BCLLS, Birmingham Ear Hospital, Annual Report, 1881.

[32] BCLLS, Birmingham Ear Hospital, Annual Reports, 1862, 1866, 1868–72, 1881–90.

[33] J. Walvin, *A Child's World: A Social History of English Childhood, 1800–1914* (Harmondsworth, 1982), 11.

institution-building. For example, children's hospitals were constructed in Dublin in 1822, Liverpool in 1851, London in 1852 and Manchester a year later.[34] The fact that children were not always treated in general hospitals, given the likelihood of their transmitting fevers into these institutions, further justified the construction of hospitals specifically intended for these young patients. In general, before the emergence of such specialist institutions, most children were treated by druggists or at home by their families. The early nineteenth century was also characterised by a very high infant mortality rate; according to statistics, a quarter of all registered deaths were those of children.[35] As many of these cases were also regarded as preventable, one could hardly make a stronger case for a specialist institution.

While all these factors were used to justify the establishment of such an institution in Birmingham, its founder, Thomas Heslop, was equally frustrated with his career in the month's leading up to the charity's foundation.[36] After working for a time in the town's leading medical institutions, and following regular quarrels with William Sands Cox at the Queen's Hospital, Heslop left the service of the town's first teaching hospital where he had been a physician since 1852 and commenced work at the less prestigious General Dispensary the following year. The Edinburgh-educated physician had first come to Birmingham in November 1848 following his appointment as medical tutor to students at the General Hospital.[37] Heslop left the service of the General in 1851 after disagreeing with staff over the organisation of the charity, evidence that suggests blame for his subsequent departure from the Queen's Hospital does not lie entirely with Cox.

As a result, though a children's hospital hardly appeared to require any justification, past conflicts with influential, albeit difficult, members of the local community encouraged Heslop to make the strongest possible case for another medical charity. Heslop began by calling a private meeting at his home on 25 June 1861; he invited thirteen influential allies, including the mayor, who arranged a public hearing of the plan at the Council House at Moor Street on 12 July. On this occasion, Heslop declared that, despite the formation of many local medical charities over the previous hundred years, none was specifically for children. Some women's hospitals had been known to take in children, but the local lying-in hospital, opened in 1842,

34 Eduard Seidler, 'An historical survey of children's hospitals', in *The Hospital in History*, ed. L. Granshaw and R. Porter (London, 1989), 185.

35 Walvin, *A Child's World*, 20–1.

36 BCLA, Children's Hospital, General Minute Book, 1861, HC/BCH/1/2/1.

37 BCLA, General Hospital, Medical Committee Minutes, 1836–55, GBH 67.

did not admit children as inpatients.[38] Neither had the town's two general hospitals in these years significantly altered their exclusionist policies towards children. In the first three months of 1861, only twelve (4 per cent) of the Queen's Hospital's 286 inpatients were children, while comprising 208 (16 per cent) of its 1,288 outpatients. At the General, figures were hardly better, children accounting for forty (8 per cent) of 499 inpatients and 541 (13 per cent) of 4,159 outpatients. Ratios were even less favourable at the town's dispensary, where only one in sixteen patients were children.[39] Poor Law medical officers, as argued by Heslop, clearly did not 'possess the confidence of the sick poor'.[40] Only by creating an institution specifically for children, he argued, would knowledge of children's diseases improve and mortality rates decrease. As at Great Ormond Street, the country's premier children's hospital, founded a decade earlier, the institution was to be open to pupils, though instruction would extend to nurses and mothers.

Following these initial meetings, Heslop's greatest opposition came from predictable quarters. Faced with similar opposition two decades earlier when organising the foundation of a teaching hospital, William Sands Cox now assumed the position of his former critics. With his own difficulties mounting, and subscriptions to the Queen's Hospital plumbing new depths, Cox rightly surmised that another hospital served only to exacerbate his problems. Consequently, he claimed there was little need for such a hospital, especially as the Queen's governors intended to open a ward specifically for children. Seeing through such arguments, the public supported Heslop's plan on the night of his initial presentation, a provisional committee having been formed before all those in attendance had left the Council House. By August, the committee's members had acquired the old eye infirmary at Steelhouse Lane and set about transforming the building into a sixteen-bed hospital for children.[41]

Although described as a specialist hospital, the proposed institution was unique in more ways than simply treating children. From the outset, and perhaps as a result of Heslop's disagreements with Cox, control over administrative matters at the hospital was placed firmly in the hands of the institution's lay administrators rather than its medical staff, as was common at other specialist hospitals.[42] The rules of the hospital also specified that

[38] J. M. Waddy, 'Report of the Birmingham Lying-In Hospital, and Dispensary for the Diseases of Women and Children', *Provincial Medical and Surgical Journal* 9, 3 (1845), 39–41.

[39] T. P. Heslop, *The Realities of Medical Attendance on the Sick Children of the Poor in Large Towns* (London, 1869), 8.

[40] Ibid., 13.

[41] Waterhouse, *Children in Hospital*, 24–5.

[42] Ibid., 26–7; BCLA, HC/BCH/1/2/1.

all medical appointments came with restricted tenures in order that openings would regularly appear, and thereby allow more than a select group of
medical practitioners the benefit of the hospital's clinical material. It was
also the town's first free hospital, a small registration fee having been introduced at its launch in order to discourage trivial cases from daily filling the
charity's waiting room. Additional funds also permitted the hospital's governors to employ a greater number of nurses than elsewhere, as staff numbers at all hospitals generally appeared to be inversely related to the age
of patients.[43] The workforce at the institution was further augmented by
a decision to establish a school for nurses alongside the hospital. In addition, as argued by Heslop, education became a large part of the institution's
remit, staff having educated many more local mothers than students during
the hospital's existence.

Compared with the ear infirmary, the growth of the Children's Hospital
was impressive, despite duplicating some of the services provided by other
existing charities. While some inhabitants might have regarded the services of the Queen's and General as sufficient to deal with the medical needs
of the town's children, in 1863, the hospital was treating almost 8,000 children, most coming from Birmingham and its adjoining neighbourhoods;[44]
the local Lying-in Hospital, by way of comparison, was treating approximately 2,000 patients annually, the majority of which continued to be
maternity cases.[45] A large amount of work would continue to be carried
out at the children's charity despite the fact that, in 1865, it still had only
twenty-two beds, a tenth of the number in use at the General that same
year, though eight more than existed at the Lying-In Hospital. In 1868,
numbers of cases at the Children's Hospital were set to expand with the
construction of a new outpatient building at Steelhouse Lane, in- and
outpatient facilities having been separated in order to prevent the spread
of infection. Interestingly, this was the same year that saw Heslop reappointed as a physician to the Queen's Hospital, following Cox's departure
from Birmingham. Two years later, this division between the Children's
Hospital's two departments was maintained when a new forty-bed hospital for inpatients was opened on Broad Street in a building that, until
the 1860s, had been the Lying-In Hospital referred to in Heslop's presentation during the summer of 1861; largely due to the cost of treatment,
the lying-in charity's governors decided to stop taking inpatients and
allowed the Children's Hospital to take over the remaining fifty-one years

---

[43] E. Lomax, *Small and Special: The Development of Hospitals for Children in
Victorian Britain* (London, 1996), 73–4.

[44] BCLA, Children's Hospital, Annual Report, 1863, HC/BCH/1/14/1.

[45] BCLA, Lying-in Hospital, Governors' Minute Book, 1842–69, MH 1/1/1.

of their existing lease for £150.[46] The new location was described as 'one of the very best sites that could possibly be selected within the borough of Birmingham for the purpose of a Children's Hospital', while the building comprised 'two large wings' and 'three magnificent Wards'.[47] Shortly afterwards, following additional expansion, the charity's capacity increased to fifty-five beds. Though late to appear and initially faced with some stiff resistance, the institution had become the town's third-largest hospital in terms of patient numbers by 1873.

As for the conditions afflicting patients, the majority of the children treated as outpatients suffered from a limited number of infectious and respiratory conditions. For example, of almost 14,000 outpatients seen in 1870, more than a third of cases were attributable to whooping cough (964), bronchitis (1,052), catarrh (765) and diarrhoea (2,683). Only 1,729 were surgical cases. Inpatients tended to represent a wider variety of diagnoses. Categorised by way of bodily systems, the majority of more than 800 cases treated in 1873 suffered from diseases of the nervous (97), digestive (113), urinary (84) and respiratory systems (77). Interestingly, although diseases of the bones and joints (86) were regularly recorded among inpatients, their numbers tended to decline with an increase in beds at the Orthopaedic Hospital in these years. A scarcity of eye and ear cases (10) also suggests staff at the Children's Hospital were avoiding duplication of hospital services by referring patients to the most appropriate specialist institutions.[48]

Despite making great efforts to separate infectious cases from other patients, staff at the hospital, as at the lying-in charity, waged a permanent battle against fevers, which regularly erupted in its wards. As a result, new measures were introduced to keep, for example, patients with scarlet fever, who comprised a tenth of inpatients in the hospital's first decade, separate from other inmates. This involved admitting children at a building across the street from the hospital proper. When found to be free of infection, children were escorted by a porter to the physician on duty, who then transferred cases to an available bed. Those children not admitted were visited at home if living nearby, otherwise mothers were ordered to call in the assistance of other medical men immediately. In some cases, mothers were permitted to report on the progress of children and collect the necessary medicines from the hospital dispensary. The hospital's medical officers also asked all mothers of infectious children by way of placards and newspaper advertisements to apply to the secretary with a note, leaving their children at home. If this request was neglected, the child was placed in the

---

[46] Ibid.

[47] BCLA, Children's Hospital, Annual Report, 1870, HC/BCH/1/14/2.

[48] Ibid.

physician's room, the parents given a notice regarding prevention of infectious diseases based on the work of Dr Budd of Bristol, who was widely recognised for his work on typhoid, or intestinal fever.[49] Usually read to parents by a nurse if they were illiterate, such regulations were freely circulated throughout the town and institution due to their limited cost, placards being hung in the hospital's waiting room, secretary's office, wards, as well as the physician's and surgeon's rooms.[50] Other boards in the waiting room concerned the general care and feeding of children. Should these precautions have failed, all inpatients were kept at least twenty-one days after the eruption of an illness, thirty days if possible, in order to prevent the spread of infectious diseases. This, among other things, further led the hospital to pioneer convalescent treatment in Birmingham.

A PLACE IN THE COUNTRY:
THE FIRST CONVALESCENT HOMES

From the founding of the General Hospital, and throughout the nineteenth century, most local hospitals received tickets from subscribers and donors which could be used to send recovering patients to convalescent homes and, in this way, free beds for more urgent cases. Most convalescent patients were sent to seaside resorts, such as Brighton and Rhyl.[51] Primarily, such destinations were seen to have two advantages: they allowed patients to be removed from towns and cities to healthier rural and seaside environments, while freeing beds for those individuals who found themselves on hospitals' earliest waiting lists. The immediate result of sending recovering patients to such retreats was a noticeable decline in the average length of time individuals remained in hospital. In general, the average stay at hospital during these years dropped from approximately thirty to twenty days. As such, the benefits of convalescent schemes immediately became clear to most hospital administrators by the second half of the nineteenth century.

From the moment they admitted their first patient in 1862, staff at the Children's Hospital had planned the establishment of their own convalescent home.[52] An offer in 1862 from Caroline Martineau, a member of the influential Unitarian manufacturing clan, resulted in the establishment of the first such home in Solihull, albeit on a much reduced scale. Nevertheless, during a single year, this small country retreat managed to free forty-two beds at the Children's Hospital for more urgent cases.[53]

49 W. Budd, 'On intestinal fever', *Lancet* 2 (1859), 4–5, 28–30, 55–6, 80–2.

50 BCLA, Children's Hospital, Annual Report, 1870, HC/BCH/1/14/2.

51 BCLA, General Hospital, Annual Report, 1862, GHB 418.

52 BCLA, Children's Hospital, Annual Report, 1861, HC/BCH/1/14/1.

53 BCLA, Children's Hospital, Annual Report, 1868, HC/BCH/1/14/2.

Shortly afterwards, drawing on her family's pin-making fortune, Louisa Ryland provided the same service to the General Hospital.[54] In the following years, numerous other small homes with between two and six beds were established in communities from Harborne to Stratford to take in convalescent patients and free beds for acute cases at Birmingham's voluntary hospitals. In 1879, the Children's received its own six-bed country cottage in Alvechurch.[55] A donkey and carriage were provided by Messrs R. and G. Tangye, the Birmingham engineers, to allow children to enjoy fresh air as well as transport them the 2 miles to and from the rail station.[56] Rail transport to Alvechurch tended to be paid for by the hospital's Samaritan Fund.

Such schemes only increased in scale over time, climaxing in 1885 with the opening of the Jaffray Hospital, the General Hospital's convalescent institution, which was built on 8 acres of land in Erdington. It, more than any previous convalescent home, resembled a building that might itself one day become an independent hospital, or, as the chairman at the Children's Hospital's annual meeting in 1870 described such homes, 'the germs … of buildings to be erected hereafter'.[57] Its foundation stone laid with Masonic honours, the £15,000 Jaffray Hospital was expected to incur a running cost of £2,500 annually. In actual fact, six years later it was costing its parent charity more than £3,200.[58] With only 310 of its own subscribers (in contrast to more than 2,000 at the parent institution) and an annual income of £770 to pay for nearly as many patients each year, most of the institution's running expense would continue to be covered by the General Hospital. As a result, in contrast to other institutions at this time, the Jaffray had a very relaxed visiting policy, being open to the general public each Monday afternoon for two hours in order to attract subscribers.[59] Though the home raised only limited funds during the remainder of the nineteenth century, the hospital survived well into the next century, eventually closing in 1991. Its founder, however, did not; John Jaffray's death was reported in 1901 in the pages of his newspaper, the *Birmingham Daily Mail*.[60]

Due to the experiments of Jaffray and the founders of the Children's Hospital, by the 1880s most other voluntary hospitals acquired similar,

[54] BCLA, General Hospital, Birmingham, Annual Report, 1863, GHB 419.

[55] BCLA, Children's Hospital, Annual Report, 1878, HC/BCH/1/14/3.

[56] BCLA, Children's Hospital, Annual Report, 1884.

[57] BCLA, Children's Hospital, Annual Report, 1870.

[58] BCLLS, General Hospital, Birmingham, Annual Report, 1891.

[59] G. Mooney and J. Reinarz, *Permeable Walls: Historical Perspectives on Hospital and Asylum Visiting* (Amsterdam, 2009).

[60] *Birmingham Mail*, 5 January 1901.

though smaller, homes. By the 1890s, even some of the smallest general hospitals in England had acquired their own convalescent homes thanks usually to one particularly generous donor.[61] A notable exception was the Queen's Hospital, which did not have its own convalescent home for the duration of the nineteenth century; as a result, patients tended to remain at the institution for longer than was common at other local hospitals. Following the success of the Jaffray, in 1891, Moseley Hall was donated to the city by Richard Cadbury, of the confectionary family, to be used as a convalescent home. The immediate benefit, however, was again realised by the Children's Hospital, as twenty of the institution's fifty beds were reserved specifically for its young patients. Consequently, the hospital's cottage in Alvechurch was closed after sixteen years, 1,383 children having each spent several weeks recovering at Copper's Hill and Arrowfield Top during this period.[62] During the next nine years, another 1700 convalescent children, many with diseased bones and joints, would be sent to Moseley Hall after being treated by the Children's Hospital's medical staff.[63]

DISEASES OF WOMEN: THE WOMEN'S HOSPITAL

While convalescent homes can be said to have been promoted and pioneered locally by the Children's Hospital, another equally innovative institution in Birmingham was the Women's Hospital. Championed by Arthur Chamberlain, and much like the Eye and Ear hospitals, it started as a small dispensary, initially located at The Crescent (now Brindley Place) in 1871, though its governors were quicker than other specialist hospitals to establish inpatient facilities, providing four free and four paying beds.[64] This decision alone made the charity very different from other voluntary hospitals, which were built to serve the 'deserving' poor only. Governors also welcomed the domestic servants of governors. Previously, besides nominal registration fees, the only paying patients received at Birmingham's medical institutions had been soldiers (See Chapter 1). In addition, though serving a broader class base than the town's other hospitals, which tended to serve only the labouring poor, its cases were restricted to those women

---

[61] Andrews, *The First Cornish Hospital*, 145.

[62] BCLA, Children's Hospital, Annual Report, 1890, HC/BCH/1/14/5.

[63] BCLA, Children's Hospital, Annual Report, 1891–1900, HC/BCH/1/14/6–7.

[64] For a good description of the hospital's founding, see J. Lockhart, 'Truly, a hospital for women': the Birmingham and Midland Hospital for Women', in *Medicine and Society in the Midlands, 1750–1950*, ed. J. Reinarz (Birmingham, 2007). See also J. Lockhart, 'Women, Health and Hospitals in Birmingham: The Birmingham and Midland Hospital for Women, 1871–1948' (PhD diss., University of Warwick, 2008).

suffering from diseases of the pelvic organs, mainly inflammations of the uterus,[65] and afflictions of the breast in special circumstances. Like many other specialist institutions, its foundation was greeted with opposition by physicians and surgeons at the town's two general hospitals, who feared that the institution would further undercut their professional authority, not to mention the financial state of their institutions.

The first medical staff appointed at the institution included a physician, Dr Thomas Savage, and three surgeons, C. J. Bracey, Ross Jordan and Lawson Tait, who was appointed junior surgeon at the age of twenty-six. In order to reduce chances of infection in a pre-antiseptic age, operations at the institution were carried out in three temporary wards or sheds, an arrangement favoured by those critical of hospitals, such as James Young Simpson, Tait's instructor while at Edinburgh.[66] Despite other risks that the charity's surgeons were prepared to take in remedying their patients' serious ailments, treatment at the institution, as elsewhere, tended to be conservative during its earliest years. In the charity's first year, staff saw only twelve inpatients, while treating 596 outpatients.[67] In general, ovariotomies were rarely attempted at any voluntary hospitals in these years, given the risks with which such procedures were associated. Additionally, as 'the "grand organs" of sexual activity in women', to members of the Victorian public the ovaries were 'the source and symbol of femininity itself'.[68] As such, their removal 'unsexed women', disrupting contemporary notions concerning women's nature and role in society. Together, these factors combined to bring this particular surgical practice into disrepute. When an ovariotomy was attempted at the General Hospital in Birmingham during the early 1850s, its necessity was not only questioned, but the incident developed into a well-publicised scandal. In this case, the surgeon, Alfred Baker, operated on a twenty-two-year old woman suffering from ovarian dropsy without consulting his colleagues. Despite the absence of the other medical staff, the whistle-blower, Thomas Gutteridge, described the 'harrowing spectacle' in a pamphlet he published privately, depicting Baker as a 'remorseless experimentalist'.[69] As a result, in subsequent years, the town's

---

[65] L. Tait, *Diseases of Women* (New York, 1879), 64.

[66] I. Loudon, *Death in Childbirth: An International Study of Maternal Care and Maternal Mortality, 1800–1950* (Oxford, 1982), 202; J. Y. Simpson, 'Report of the Edinburgh Royal Maternity Hospital, St John's Street', *Monthly Journal of Medical Science* 9 (1848–9), 329–38.

[67] BCLA, Birmingham and Midland Hospital for Women, Annual Report, 1872.

[68] O. Moscucci, *The Science of Women: Gynaecology and Gender in England, 1800–1929* (Cambridge, 1990), 134.

[69] T. Gutteridge, *The Crisis: Another Warning Addressed to the Governors* [*General Hospital, Birmingham*] (Birmingham, 1851), 19–20.

two general hospitals tended to confine themselves to maternity cases. Had they not done so, their death rates would inevitably have escalated, circumstances that would undoubtedly have affected subscription levels, as well as eliciting repeat protests from Gutteridge.

At the Women's Hospital, on the other hand, ovariotomies were common practice. As elsewhere, ovariotomy in certain medical circles became accepted long before any significant diminution in mortality rates had been achieved.[70] Convinced by his mentor, James Young Simpson, that abdominal surgery was both feasible and justifiable, Lawson Tait conducted his first ovariotomies soon after he was appointed to his first surgical position at the Clayton Hospital in Wakefield in 1867.[71] Very daringly, he continued to perform the procedure, described as 'belly ripping' in some medical circles,[72] following his move to Birmingham in 1870, though with little noticeable success. The death rate for such cases at the Birmingham Women's Hospital between 1871 and 1877 ranged between 30 and 100 per cent.[73] Though one might have expected the charity's popularity to suffer, the staff's openness with even their poorest results helped them retain a certain amount of public support and allowed its surgeons, in time, to break new ground in their respective specialist field. According to Tait, for too long medical practitioners' prime concern had been with women's reproductive functions. 'The subsidiary relations of her special organs and the special acquirements of her physique, based upon these, [had] necessitated the establishment of another class of specialist, the gynaecologist.'[74] In Birmingham, Tait discovered a free-thinking environment that was conducive to both religious dissent and innovation in medicine.[75]

Given the serious nature of its cases, the charity remained equally popular with students, though their numbers were strictly controlled. Soon after the hospital had opened, its medical board decided to admit students as long as they had successfully completed their third winter session, had passed their first professional exams and presented a certificate attesting to their having previously filled the office of clinical clerk or surgical dresser.[76] Neither were students charged a fee, unless a certificate of attendance was

---

[70] Moscucci, *Science of Women*, 151.

[71] W. J. S. McKay, *Lawson Tait: His Life and Work* (London, 1922), 18–19.

[72] J. E. Sewell, 'Bountiful Bodies: Spencer Wells, Lawson Tait, and the Birth of British Gynaecology' (PhD diss., Johns Hopkins University, 1990), 116.

[73] BCLLS, Women's Hospital, Annual Reports, 1871–7.

[74] R. L. Tait, *Diseases of Women and Abdominal Surgery* (Leicester, 1889), 3.

[75] Sewell, 'Bountiful Bodies', 159–60, 176, 181, 324–6.

[76] BCLA, Birmingham and Midland Hospital for Women, Medical Board, 1871–92, HC/WH/1/5/1.

1 John Ash (1722–98), portrait of 1788 by Joshua Reynolds.
The General Hospital is in the background; Ash is holding the plans.

2 Richard Middlemore, surgeon and
leading ophthalmologist

3 William Sands Cox (1802–75), surgeon
and founder of the Queen's Hospital

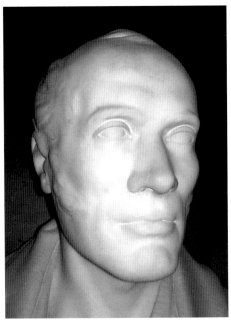

4 Gabriel DeLys (c. 1783–1831), a bust in
St Bartholomew's Church, Edgbaston

5 Annie Clarke, physician at the
Children's and Women's hospitals

6 Langston Parker (*c.* 1805–1871), surgeon to the Queen's Hospital, local syphilis expert and father of Samuel Adams Parker

7 Lawson Tait (1845–99), surgeon to the Women's Hospital

8 Birmingham workhouse (built 1733) and its infirmary wing (completed 1766)

9 General Hospital, drawn and engraved by T. Radclyffe, *c.* 1820

10 Birmingham General Dispensary, *c.* 1820

11 Queen's College from Ackermann's Panoramic View of Birmingham, 1847.

12 New museum of Queen's College Medical School, *c.* 1856

13 Queen's Hospital and its new outpatients department,
paid for by the Birmingham Saturday Fund, 1878

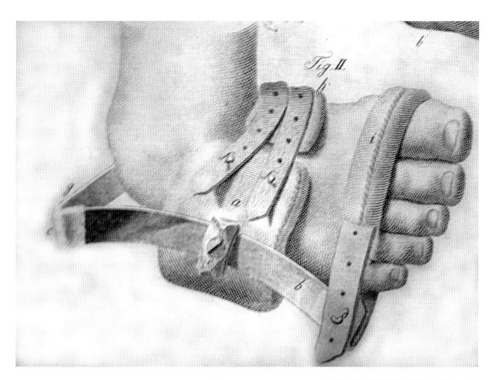

14 *(above)* Child's club foot; engraving from A. Scarpa's *A Concise Treatise on Dislocations and Fractures*, 1835

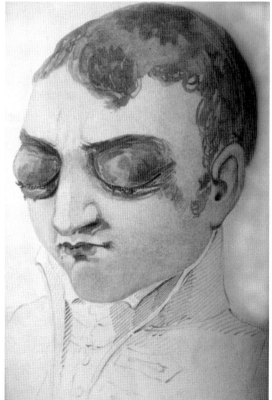

15 *(right)* A case of Egyptian ophthalmia; from J. Vetch's *Diseases of the Eye*, 1820

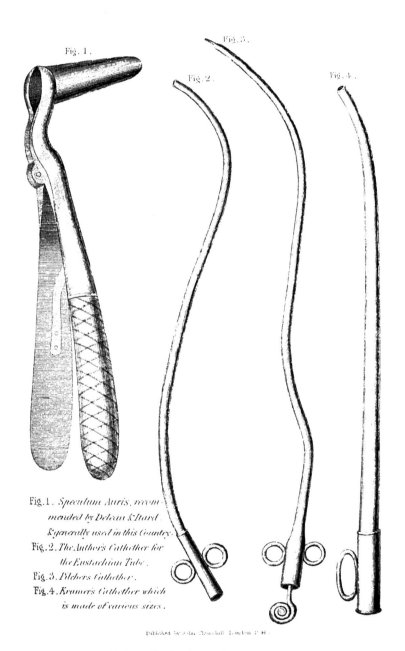

Fig. 1.

Fig. 3.

Fig. 2.

Fig. 4.

Fig.1. *Speculum Auris, recom-*
mended by Deleau & Itard,
& generally used in this Country.

Fig.2. *The Author's Cathether for*
the Eustachian Tube.

Fig.3. *Pilchers Cathether.*

Fig.4. *Kramer's Cathether which*
is made of various sizes.

Published by John Churchill, London P.H.

16 William Dufton's aural instruments;
from his *The Nature and Treatment of Deafness and Diseases of the Ear*, 1844.

17 7 The Crescent, site of the Women's Hospital, 1871–8,
and Lawson Tait's private hospital, 1882–95

18 Skin and Lock Hospital, John Bright Street, 1886

19 Women's Hospital, Stratford Road, Sparkhill, *c.* 1890

20 Ear Hospital, Edmund Street, 1896

21 Ward at the Children's Hospital, *c.* 1890

22 Children's ward at the Orthopaedic Hospital, Newhall Street, *c.* 1898;
photograph by J. Hall Edwards, radiologist

required, in which case £2 2s was to be paid to the hospital secretary. Other conditions limited their numbers to no more than two students in the out-patient department at once, and any students visiting the wards were to be accompanied by a medical officer. Expressing themselves against the crowding of medical gentlemen, the hospital committee claimed to respect patients' objections to the presence of spectators during examinations or operations. Equally, 'every precaution [was] taken to avoid injury to the most delicate feelings', women being seen one by one in the presence of a nurse and in beds affixed with curtains, at least until June 1875.[77] Similarly strict regulations were introduced at the Samaritan Free Hospital for Women and Children in London, where Spencer Wells had begun to carve out a reputation in gynaecology two decades earlier.[78]

Despite initially reporting poor results, the hospital's governing board soon acquired the funds to extend the charity. Initially, the governors pur-chased the house next door to their original building in order to separate in- and outpatients, as at the Children's Hospital. Outpatients, who num-bered between 80 and 120, were each charged a fee, which was made nomi-nal so as not to deter any serious cases, discouraging only those 'brought by curiosity or some motive other than a real desire to place themselves under the care of the Medical Officers'.[79] Nevertheless, staff experienced some difficulties ensuring the regular attendance of outpatients. Of 400 outpatients, forty-six attended only once, forty-two twice, in which case eighty-eight 'failed to become patients in reality'.[80] According to the hos-pital medical staff, who clearly used their posts to undertake some early clinical research, time spent on such cases without the benefit of follow-up appointments was 'utterly wasted'.

Increased work soon led staff to restrict the number of free cases. Of eight-one patients attending in 1872, sixty-eight were treated free of charge, while thirteen were paying patients.[81] By 1875, total numbers rose to eight-one (thirty-one paying and fifty free), despite the existence of only seven beds.[82] Most serious operations were now performed by Tait in a wooden hut, measuring 23 × 17 feet. Of thirty patients undergoing major operations in 1875, 50 per cent died. Outpatients had also increased, reaching 1,785 women, who made a total of 11,325 attendances. As a result, each surgeon

[77] Ibid.

[78] Sewell, 'Bountiful Bodies', 93–4.

[79] BCLA, Birmingham and Midland Hospital for Women, Medical Board, 1871–92, HC/WH/1/5/1.

[80] Ibid.

[81] Ibid.

[82] Ibid.

was to appoint an assistant to administer chloroform at operations and otherwise help attend the rising number of outpatients. The presence of an assistant was an additional insurance against charges of improper conduct during such procedures, as, according to Lawson Tait, 'even in the minds of the purest women … delusions occur during the anaesthetic condition which retain strong hold of their waking moments'.[83] From 1872, patients were also required to pay towards the more than 300 therapeutic instruments which the hospital distributed at a cost of approximately £20 annually.[84] However, payment was often waived when made awkward. For example, unlike splints and trusses, many of these gynaecological instruments, including intra-uterine, cervical and vaginal supports, were applied during operations without the knowledge of the patients. Consequently, staff decided instead to raise the registration fee to a shilling in order to cover the additional expenses associated with treatment.

While such charges may have covered the costs of surgery, the hospital premises remained unsatisfactory to staff. Although the hospital's consulting room had originally been regarded as 'the best of any women's hospital in Britain', by the summer of 1873 staff no longer regarded it as satisfactory.[85] Besides being poorly lighted, it had no special arrangements for ventilation and was of a small size. As a result, the air in it quickly became impure, 'so much so as to be very distressing to the Staff and their assistants and also very unhealthy'.[86] Though any offensive cases, as at many other institutions, were moved to a separate ward in the upper story of the hospital, by April the following year, the governors were already discussing a new building. When asked about their requirements, surgeons claimed the outpatient department should comprise two consulting rooms, a private room with WC, a room for a nurse to register cases, a waiting room with capacity for 150 patients and a dispensary. Moreover, the outpatient department was to be central and near transport routes. Inpatient facilities, in contrast, were not to be centrally located, but surrounded by open ground with room for expansion. Staff examined a house on The Crescent and a ward at the local Homeopathic Hospital (founded 1845) at Upper Priory for the purposes of inpatients, the premises at no. 11 The Crescent being regarded as superior. Instead of moving, three more wards were erected in the garden of the nearby no. 8. However, the first set of operations performed in these sheds met with disappointing results, which staff appear to have associated

---

[83] Tait, *Diseases of Women*, 34.

[84] BCLA, Birmingham and Midland Hospital for Women, Medical Board, 1871–92, HC/WH/1/5/1.

[85] Ibid.

[86] Ibid.

with the number of outpatients who had been admitted to the hospital's six wards.[87] Operations continued, but it was decided that the sheds be used for ovariotomy cases only, a set of guidelines being additionally drawn up for the nurses. Determined to ascertain the cause of each death at the hospital, a pathologist, Robert Saundby, was also appointed. Besides helping administer anaesthetics, Saundby performed an autopsy on each case dying in hospital, staff being required to attend all such sessions.

By February 1878, following a donation from Louisa Ryland, the hospital board announced their decision to purchase another, larger building in Sparkhill. In 1878, an enlarged hospital with twenty-one beds was opened, while outpatients attended a clinic next to the Children's Hospital, which had been erected at Upper Priory in the same year. A year later, its medical officers recorded treating 162 inpatients, who were now charged a registration fee of four shillings; 236 were seen in 1880.[88] Outpatients, on the other hand, averaged 2,000. Additionally, an electrical circuit was placed between Sparkhill and Upper Priory, allowing Dr Savage and Mr Tait to be informed of emergencies more promptly. Soon after, the surgeons' homes too were connected to the hospital by telephone; Tait lived at no. 7 The Crescent. Given such technological developments, not to mention the fact that both surgeons treated a number of their private patients at the institution, the presence of senior medical staff at the hospital was noticed to a greater extent than at other voluntary hospitals. By 1882, however, Tait acquired the premises at no. 8 The Crescent, which he transformed into his own private hospital.[89]

Unlike the town's other medical charities, the Women's would not establish its reputation as a result of the numbers of patients cured or relieved alone. Its status was based entirely on other achievements. As at its first building, four cottage wards were erected in the new hospital's garden, where the charity's surgeons, Tait and Savage, continued to perform and perfect ovariotomies on charity and paying patients. Despite their initial failures, as in previous years, staff made efforts to determine the reasons for their poor results. At first, ovariotomies were restricted to one a day. On another occasion, Tait requested Alfred Hill, Birmingham's Medical Officer of Health, to analyse hospital water samples. When these were found to be contaminated with animal impurities, alternative sources were sought. Though strict cleanliness was the rule, staff still used considerable amounts of carbolic, or phenol to disinfect wounds and instruments; ether,

[87] Ibid.

[88] BCLA, Birmingham and Midland Hospital for Women, Medical Committee Minutes, 1871–92, HC/WH/1/5/1.

[89] Sewell, 'Bountiful Bodies', 196.

on the other hand, had become the preferred anaesthetic as it was less likely to induce nausea in patients. As a result, death rates following the risky procedure of ovariotomy rapidly dropped from 20 per cent to 9 per cent in 1881, when fifty-eight operations were conducted.[90] Surprisingly, the latter figure was also the hospital's average death rate in these years, despite the performance of more than 120 abdominal sections annually. In the process, medical staff won over some of their staunchest critics, such as Oliver Pemberton, a surgeon to the General Hospital who, as a student, had attended William Sands Cox's original anatomy lectures in 1825 before himself becoming a lecturer in anatomy at Queen's College, Birmingham. Pemberton, who had originally doubted the need for the Women's Hospital, later regarded it as an important teaching institution.[91]

In 1881, the hospital, like the Children's nearly two decades earlier, and the Queen's Hospital in 1875, was also made free, even though expenses, approaching £1,646, exceeded income. Receipts that year totalled £1,291, £525 (41 per cent) coming from subscriptions, another £145 (11 per cent) from Hospital Saturday, £405 (31 per cent) from registration and paying patients, £32 (2 per cent) from the sale of instruments and other materials, and the remaining sums being derived from various popular entertainments. Compared with the Women's Hospital, the Children's Hospital raised £3,200, of which subscriptions comprised 50 per cent and registration fees only 17 per cent. Besides receiving twice as much from the Hospital Saturday Fund, which was distributed according to patient numbers, the Children's Hospital also obtained 7 per cent of its income from investments. In contrast, the governors of the Women's Hospital could not afford to invest any of the charity's income in these years. Nevertheless, until such predictable sources of revenue increased in an age when subscriptions were of declining importance to overall earnings, the fortunes of voluntary hospitals could fluctuate wildly with donations and legacies. Consequently, though not appearing in a position to extend services at the end of 1881, the hospital committee, like that at the Children's more than a decade earlier, proposed the establishment of a convalescent fund the following year to which an anonymous lady, almost certainly Louisa Ryland, offered £1,000.[92] Four years later, in 1886, a similar act of generosity led Lady Lambert of Enville to offer the hospital a four-bed cottage and the services of a nurse to care for convalescent patients.[93]

This particular year became an equally significant milestone for the

---

[90] BCLA, Women's Hospital, Annual Report, 1881.

[91] Ibid.

[92] BCLA, Women's Hospital, Annual Report, 1882.

[93] BCLA, Women's Hospital, Annual Report, 1886.

hospital's ordinary inpatients. After the relapse of a patient in 1886 following an operation, it was decided by medical staff not to release any inpatients in the first six days following any surgical procedure.[94] A number of other precautionary measures seen at other institutions, only implemented more comprehensively at the Women's Hospital, were subsequently introduced. In general, access to patients was restricted after this date. To begin with, the managing committee revised the hospital's regulations, ruling that no student be admitted to the inpatient department.[95] Only one member of staff, on the other hand, was permitted to enter the outpatient department in Upper Priory at a time. Furthermore, their attendance at operations was permitted only after having signed in at the inpatient department and if their presence in no way endangered the patients. In such cases, forty-eight hours were to pass between each major operation before admitting the next patient to theatre in order to allow sufficient time to clean and disinfect the facilities. In case of death, a ward was to remain empty for four days or more. Presence of staff at post-mortems, though once deemed essential, was no longer required, as each was engaged in private practice and their presence at post-mortems would be 'objectionable' to their private patients. When it was suggested that only one major operation be performed each day to prevent further danger to patients, as well as strain the nursing staff, however, the medical board decided it was totally acceptable and less taxing to perform two major operations a day, the average ovariotomy having lasted between five and six minutes, while preparing a ward for such an operation took up to four hours.[96] Not only was it found to be less straining than scheduling operations on two consecutive days, but Tait claimed he carried out three major operations a day at his private hospital at The Crescent.[97] Over the following weeks, other reforms were suggested. Most notably, the yard ward was finally judged unsuitable and was transformed into a more substantial building, raised completely from the soil. So, too, were the drains of the main building adequately trapped and all patients' bandages burned. Given that the list of precautionary measures grew longer with each meeting, plans for a new hospital were soon submitted to the governors.

Reporting more positive recovery rates than even the most conservative of provincial voluntary hospitals, specialist or otherwise, staff went on

---

[94] BCLA, Birmingham and Midland Hospital for Women, Medical Committee Minutes, 1871–92, HC/WH/1/5/1.

[95] Ibid.

[96] Tait, *Diseases of Women*, 169, 172.

[97] BCLA, Birmingham and Midland Hospital for Women, Medical Committee Minutes, 1871–92, HC/WH/1/5/1.

to acquire an international reputation over the next decade. Perhaps most surprisingly, the hospital's death rate continued its decline, reaching 1.4 per cent in 1892.[98] Often overshadowed by his younger colleague at the hospital, Thomas Savage had a success rate that rivalled that reported by Tait. According to his publications, and in contrast to Lister's first writings on antiseptic surgery, the secret of their success was short incisions, which prevented interference with organs other than those to be removed, and thereby ensuring better drainage, as well as restricting the 'door opened for the introduction and absorption of septicity'.[99] Many published case studies also reveal, however, that surgeons' concerns for cleanliness were carried to extremes, if not combined with antiseptic methods.[100]

This was not, however, the sole reason for the hospital's reputation. Besides pioneering ovariotomies, the institution's staff also helped expand women's roles in medicine locally, if not nationally. Of all Birmingham's hospitals, the Women's was the first to appoint a female practitioner, Zurich-educated Louisa Atkins having been elected House Surgeon over two male candidates in 1872. The following year, the Medical Committee also permitted her to take a female pupil.[101] The fact that she was to be instructed in medical science, as well as the duties of a nurse, suggest women had not yet attained the status of their male colleagues. Nevertheless, the experiment, simultaneously regarded a good advertisement for the charity, was judged a success, and Mary Pechey succeeded Atkins to the post of House Surgeon in 1875, when the latter left Birmingham to join Elizabeth Garrett Anderson on the staff of the Hospital for Women in London.

The Children's soon followed this example with the appointment of Anne Barker and Annie Clarke to its resident staff in 1878 and 1880 respectively, Clarke evidently having possessed a more complete scientific knowledge than her male rival to the post.[102] In 1893, Dr Clark left her post at the Children's when she was elected physician to the Women's Hospital. Far from being a local phenomenon, the entry of women into more responsible positions would become a common pattern at women's and children's hospitals throughout Europe in these years.[103] However, the

[98] BCLA, Women's Hospital, Annual Report, HC/WH/1/10/4.

[99] T. Savage, 'On removal of the uterine appendages', *British Medical Journal* (8 Jan 1887), 52; Tait, *Diseases of Women*, 171.

[100] Tait, *Diseases of Women*, 168–70.

[101] BCLA, Birmingham and Midland Hospital for Women, Medical Board Minutes, 1871–92, HC/WH/1/5/1.

[102] Tait, *Diseases of Women*, 139.

[103] See, for example, H. Marland, '"Pioneer work on all sides": the first generations of women physicians in the Netherlands, 1879–1930', *Journal of the History of Medicine and Allied Sciences* 50, 4 (1995), 441–77.

Women's Hospital also stands out as the first local hospital to appoint a female dispenser, Miss Harding having been appointed to this post in 1873, largely because this proved to be cheaper than appointing a male dispenser. Although cheaper, her instructor, Mr Lucas of Colmore Row, claimed he never had a pupil who was 'more attentive', or 'one who more quickly acquired proficiency'.[104] Again, the Children's followed the Women's example in 1877 by appointing a female dispenser, in this case a former pupil at the Women's Hospital.[105] The Orthopaedic Hospital made a similar appointment in 1891.[106] Though not the first English hospitals to employ female dispensers, by this time, hospitals in Birmingham offered opportunities equal to, if not greater than, those available in London.[107] Finally, the Women's Hospital's governors also broke new ground by appointing women to managerial positions at the institution. In fact, an early board even introduced a rule stipulating that half of the eighteen-member managing board be women. Appropriately, the hospital also became the first local medical charity to appoint a female President when the Countess of Dudley commenced her presidency at the hospital in 1893. Coincidentally, this was the same year female factory inspectors were also appointed nationally. While women were making notable professional gains throughout England at this time, most provincial voluntary hospitals witnessed such developments during the interwar period.

Surprisingly, and in great contrast to the charity's other ground-breaking work, the nurses at the Women's Hospital were the last at any local voluntary hospital to be offered any systematic form of training. Though having commenced their work by employing nurses from the town's Nurse Training Institution, by 1876, staff had decided that an 'ordinary intelligent woman' could undertake outpatient work as efficiently as a trained nurse and at much less cost, a fact that Tait emphasised in his publications.[108] In 1883, ideas had changed, and the hospital received a letter from Arthur Chamberlain suggesting that nurses be given lectures as at other local hospitals. However, the Medical Board declined an offer from Dr Nelson, who proposed to deliver a series of lectures to the nurses, preferring instead that they attend those of the St John's Ambulance Association or even those

[104] BCLA, Birmingham and Midland Hospital for Women, Medical Board, June 1871 – March 1892, HC/WH/1/5/1.

[105] BCLA, Children's Hospital, Annual Report, 1877, HC/BCH/1/14/3.

[106] BCLA, Orthopaedic Hospital, Medical Committee Minutes, 1884–1903, HC/RO/Box 9.

[107] E. Jordan, '"Suitable and remunerative employment": the feminization of hospital dispensing in late-nineteenth-century England', *Social History of Medicine* 15, 3 (2002), 441.

[108] Tait, *Diseases of Women*, 167.

offered to staff of the Queen's Hospital; the board did not consider a special course to be necessary, given the small number of nurses.[109] Three months later, sensing the error in their decision, the medical committee accepted Nelson's offer. When an outbreak of infection six years later, in 1889, pointed to nurses as carriers of infection, the surgeons finally and unanimously decided that a better quality of nurse was required at Sparkhill. As a result, the hospital's nurses were released from their menial work, similar reforms having been carried out at other hospitals decades earlier. This episode, above all, serves to remind us that, while some of Birmingham's teaching hospitals might have been quicker to establish international reputations in one particular area, each institution developed at its own particular rate. Usually this was a question of finance, other times the result of individual initiative, or public support. That said, innovation in one field, never guaranteed the adoption of best practice in all other areas of hospital activity.

[109] BCLA, Birmingham and Midland Hospital for Women, Medical Board, June 1871 – March 1892, HC/WH/1/5/1.

CHAPTER 5

The Importance of Good Teeth and Skin

T HOUGH FOUNDED before the Children's and Women's hospitals, the history of Birmingham's Dental Hospital sits more comfortably beside that of the Skin Hospital than other local specialist institutions. Issues of chronology aside, these two institutions share a number of similarities and a common trajectory in their first decades of existence and will therefore be considered jointly in this chapter. To begin with, both of these medical specialties, dentistry and venereology, were for centuries very closely linked to quackery. As such, this chapter additionally offers an example of the way in which two emerging medical specialisms professionalised in the second half of the nineteenth century. Given this shared heritage, medical practitioners associated with these charities often very consciously defined themselves in opposition to their quack forefathers. For example, the founder of the first dental charity in Birmingham regularly contrasted his very aggressive dental techniques with the less invasive and, therefore, ineffectual methods of charlatans and mountebanks, thereby demonstrating the confidence of a new generation of English dental practitioners. However, despite such efforts, many links with these practitioners' disreputable pasts remained. For example, both institutions were very much reliant on payments from patients for their economic survival, and publications in both fields often continued to read very much like quack advertisements, not scientific texts. Nevertheless, a process of change was very clearly under way, and, by century's close, these two areas of medical specialisation had been largely transformed. This is perhaps best demonstrated by the new status achieved by both the Dental and Skin hospitals, which were officially recognised by Birmingham medical school as teaching hospitals in the last decade of the nineteenth century.

SCALING NEW HEIGHTS:
THE DENTAL HOSPITAL

Though appearing very different from Birmingham's other voluntary hospitals, the Dental Hospital shared certain similarities with those institutions whose histories have already been outlined. Established by Samuel Adams Parker in January 1858, nearly a year before a comparable institution was opened in London at Soho Square, its founder equally shared some important similarities with William Sands Cox, the founder of

Birmingham's medical school.[1] Not only did both men initially set up their respective institutions at the age of twenty-seven, but, like Cox, Parker relied on the support of some of the town's senior practitioners when first establishing his medical institution. His father, Langston Parker, was instrumental in this respect, having formally served as a surgeon at the Queen's Hospital.[2] Also, while Cox had studied with many well-known London practitioners, including Astley Cooper, Parker studied for a time at the Middlesex Hospital in the capital with the nation's premier dental surgeon, John Tomes, regarded by many as the father of modern dentistry.[3] Moreover, the hospital, like those previously discussed, originally opened as a dispensary, moved often during its early years and expanded only cautiously.

In many ways, Parker's timing, like that of Cox, could not have been better, these being equally important years in the development of the dental profession. While the first meeting of the British Odontological Society had taken place in January 1857, the Royal College of Surgeons was granted its dental charter in 1859, awarding its first Licentiate in Dental Surgery the following year. Thereafter, few hospitals in the United Kingdom would not have 'a legally qualified dentist upon their medical staff'.[4] Interestingly, the LDS was originally proposed by J. L. Levison of Birmingham in a letter to the *Lancet* in 1841.[5] Twenty years later, the first dentist in the town had acquired the new degree, Parker having travelled to London in 1861 to obtain his qualification.

The fact that the founder of the Dental Hospital's only publications dealt with the manufacture and fitting of false teeth suggests preventive dental medicine took some time to emerge in Birmingham, let alone other provincial centres in these years. Even before the foundation of Parker's charity, however, a limited amount of dental care had been available to Birmingham's poor. More importantly, these services were not limited to those resulting from the initiative of innovative itinerants. The surgeons Thomas Clarke and Thomas Tomlinson, for example, both appear to have offered dental treatment in Birmingham during the

---

[1] E. Smith and B. Cottell, *A History of the Royal Dental Hospital of London and School of Dental Surgery, 1858–1985* (London, 1997), 5.

[2] A. F. Stammers, 'The Birmingham Dental Hospital and School', *British Dental Journal* 105, 3 (1958), 77.

[3] R. Cohen, 'The Birmingham Dental Hospital: with some account of the founding of the Dental School', *Birmingham Medical Review* 20 (1958), 331; Smith and Cottell, *History of the Royal Dental Hospital*, 3.

[4] S. A. Parker, *Remarks upon Artificial Teeth* (Birmingham, 1862), 56.

[5] E. G. Forbes, 'The professionalization of dentistry in the United Kingdom', *Medical History* 29 (1985), 169.

1790s.[6] Additionally, since 1822, William Robertson, who made an important early contribution to the understanding of caries, is to have attended gratuitously 'any of the poorer classes who bring recommendatory notes from respectable persons' at his practice in the Old Square.[7] According to Robertson, in 1846, the 'dentist ... is generally to be found only in larger towns',[8] and, by the 1850s, directories for the town record sixteen dental practitioners of varying status in Birmingham. Additionally, local papers also advertised an array of products and services from dentifrices to dentures available from both travelling dentists and local druggists and chemists.

Dental care was also offered at some of the town's other hospitals, though their work was rarely of a preventive nature. According to the General Hospital's annual reports, for example, in 1837, more than 1,600 teeth had been extracted by surgeons and students at the town's main voluntary medical institution. Among many other Birmingham graduates, the Victorian polymath and eugenicist Francis Galton developed considerable proficiency at drawing teeth while a pupil at the General Hospital in these years.[9] By 1864, some six years after the Dental Hospital was founded, the number of extractions had more than doubled, staff having removed approximately 3,500 teeth from the infected mouths of Birmingham's inhabitants, most being pulled by students reputedly under the supervision of the General's house surgeon, Charles Bracey; by way of comparison, a similar number of dental cases were treated at medical charities in Newcastle thirty years later.[10]

Such dental services were not confined to Birmingham's largest medical charity either, as is perhaps best suggested by the first gifts to the Children's Hospital, all of which were listed in the charity's annual report for 1861. Donations in this year included a scale, clocks, a set of ward and bath thermometers and various surgical instruments, including a complete set of dental tools. Surely stimulated by the latter gift, in December 1861, the hospital board instructed one of their surgeons, Mr Howkins, to organise a dental department at the institution. Though only four 'teeth

[6] *Aris's Birmingham Gazette*, 22 July 1793; C. Hillam, *Brass Plate and Brazen Impudence: Dental Practice in the Provinces, 1755–1855* (Liverpool, 1991), 7–8, 18.

[7] R. A. Cohen, E. A. Marsland and C. Hillam, 'William Robertson of Birmingham 1794–1870', *British Dental Journal* 142 (1977), 65; R. Cohen, 'English theories of the causes of dental caries, 1800–1850', *Dental Magazine and Oral Topics* 53, 6 and 7 (1936), 10.

[8] W. Robertson, *Practical Observations on the Teeth* (London, 1846), 202.

[9] F. Galton, *Memories of My Life* (London, 1908), 22–47.

[10] J. J. Murray, I. D. Murray and B. Hill, *Newcastle Dental School and Hospital, 1895–1995* (Newcastle, 1995), 4.

cases' were seen by hospital staff in 1861, four alone were treated in the first month of the following year. While this suggests a gradual speciali-sation in the hospital's work, which now included considerable bone and joint work, as well as a significant number of skin cases,[11] little additional dental equipment was purchased over the next decade. Only in 1873 was a chair specifically for dental operations acquired by hospital staff. As the slow growth of this particular department perhaps suggests, the existence of dental departments at other Birmingham hospitals was never a serious impediment to the growth of a hospital intended solely for dental cases.

Parker's dental dispensary was first located at 13 Temple Street in the town's Oddfellows Hall. Unlike the dental department at the Children's, it admitted 645 patients in its first year, an almost equal number of men and women having attended the charity. Parker, the hospital's dental surgeon, performed 725 operations, including 479 extractions, most being conducted between 9 and 10 am on a single morning each week.[12] Patients quickly surpassed 1,000 the following year, the number of operations practically mirroring increases in patients. Operations, however, doubled by 1860, reaching 2,921, while the total number of cases was 1,638, figures which suggest some patients attended for multiple surgical procedures. For the next five years, patients averaged 2,500, outnumbered only slightly by the total number of operations. Similar increases would be recorded at den-tal charities in other provincial towns in these years. The next two provin-cial dental dispensaries appeared in Liverpool and Plymouth in 1861, the former having been modelled closely on Parker's institution.[13]

In 1863, the Birmingham charity moved to 2 Upper Priory, where it shared premises with the Homeopathic Hospital, a suitable pairing given homeopathy's equally suspect standing in these years.[14] If anything, the reputation of dentistry at this time differentiated the venture most from other medical charities and perhaps made Parker even more cautious to expand his charity, given that his chosen speciality was the least reputable of the recognised medical specialties. Throughout much of the nineteenth century, dentistry had been associated with and was practised by quacks, itinerants, and, perhaps most detrimental of all in the eyes of the public, women.[15] Like local trades and businesses, dentists also advertised more than practitioners in other medical fields. In general, the medical profession

---

[11] BCLA, Children's Hospital, Annual Report, 1870, HC/BCH/1/14/2.

[12] *Birmingham Daily Post*, 12 January 1859.

[13] E. Muriel Spencer, 'Notes on the history of dental dispensaries', *Medical History* 26 (1982), 55.

[14] *Lancet*, 20 February 1875; 13 March 1875.

[15] Hillam, *Brass Plate and Brazen Impudence*, 37.

discouraged orthodox practitioners from advertising for patients and, as a result, the practice of heavy advertising became even more closely associated with quackery. This alone would explain why a decision to devote nearly half of the dental dispensary's income, or £120, towards advertising the charity soon after its foundation was regarded as suspect in some quarters of the community.[16] Most hospitals in these years did not spend more than a tenth of this sum on advertising their work. Nevertheless, despite the reluctance of other hospitals to follow suit, with an average of 3,000 cases being treated at the institution in these years, few could argue that Parker's strategy had failed.

Despite printing many ads, notices and hospital reports, undoubtedly Parker's best advertisement in these years were his publications on the subject of artificial teeth. Though many other dental texts had appeared before Parker's, the dental trade itself had only recently benefited from the appearance of its own periodical, the first issue of the *British Journal of Dental Science* dating from 1856. Should there have been any doubts concerning Parker's medical credentials in these years, a glance at his captivating treatise, *Remarks upon Artificial Teeth and upon the states of the mouth in which they should and should not be used* (1862), would have convinced even his critics that this young Birmingham dentist belonged to a new elite in dental surgery. To begin with, his short, but intriguing, book is dedicated to John Tomes, which alone would have lent it considerable credibility, Tomes having produced his very practical and popular tome, *A System of Dental Surgery* (1859), three years earlier. Unlike Tome's work, which includes no actual details of the patients he treated while serving as dental surgeon to the Middlesex Hospital, Parker's is clearly based on specific cases attended at his dental dispensary in its first years. Given that the Birmingham charity's earliest printed annual report appears to have been issued in 1879, the contents of this titillating text are particularly interesting.

Like the establishment of his dental charity, the timing of Parker's publication could scarcely have been better, vulcanite, or gutta percha, having revolutionised prosthetic dentistry in 1859. Before the introduction of vulcanised India rubber, artificial teeth in England were often made of ivory, that from the hippopotamus in the best and costliest prostheses, the softer and noticeably inferior bone of the walrus tusk more generally.[17] Walrus teeth became soft and rotten after only twelve months of use; those of the hippopotamus reputedly lasted some three or four years.[18] In both cases, teeth eventually turned yellow, then blue and made the breath of the wearer

[16] Cohen, 'The Birmingham Dental Hospital', 332.

[17] N. Parks, 'Dentistry of fifty years ago', *The Dental Record* 22, 11 (1902), 488.

[18] Parker, *Remarks upon Artificial Teeth*, 42.

very offensive, surely as foul of that of the average walrus, even if washed half a dozen times a day. In contrast, vulcanised sets of artificial teeth were easier to keep clean, weighed less than ivory teeth, and were said to be nearly indestructible. They also allowed dentures to be made in a colour much closer to that of natural gum.[19] Such characteristics were unheard of in previous eras and, not surprisingly, impressed a new toothless generation, who claimed the teeth were so good, they actually forgot they had them in their mouths.[20]

Of equal importance to this innovation, if not greater according to some dentists, was the introduction of inclining dental chairs in 1858.[21] Previous to this development, most dental operations were conducted in ordinary armchairs. To compensate for the poor positioning of patients, some chairs had been affixed with neck-level bars which provided some extra leverage for extractions, but an increase in fractured teeth is said to have discouraged such harsh methods and instead encouraged the introduction of special forceps for individual teeth. Either way, such earlier practices probably only helped Parker build up his private practice and finance his charitable work.

In general, most dentists designed their own instruments. Though anaesthetics may have spared numerous patients additional torture, many dental practices were clearly low-tech and artisanal in nature. Most cavities, for example, were cut out solely with hand drills, excavators and files. Ice was used more often than anaesthetics to dull the pain of these operations.[22] Cut surfaces on teeth were treated with creosote and wool until the pain ceased, when fillings would be inserted on top of dressings. Bibulous paper was used instead of rubber dams to keep cavities dry; precipitated silver and Sullivan's amalgam were the preferred material for fillings, though yellow sealing wax, or even bee's wax, was occasionally used without success.[23] Though differing little from the materials of the 'quack', most dentists endeavoured to appear very different from the average charlatan. For example, while the questionable practitioner was to have made a performance of an extraction, under no circumstances was the professional dentist to display his instruments.[24] While the publications of Tomes, and many others, no doubt taught a new generation of dental surgeons to speak

[19] R. King, *The History of Dentistry: Technique and Demand* (Cambridge, 1997), 23.

[20] Ibid., 4.

[21] Interview with Professor Ronald Cohen, 6 September 2000.

[22] Senex, 'A reminiscence of dentistry in the old days: a fortunate dental student, 1855', *British Journal of Dental Science* 68, 1225 (1920), 9.

[23] John Gray, *Dental Practice* (London, 1837), 24.

[24] Parks, 'Dentistry of fifty years ago', 498.

like men of science, primarily by introducing them to specific anatomical and pathological terms and expressions, early texts on dental ethics suggest the respectable dental practitioner of mid-century took 'no snuff and eschew[ed] tobacco'.[25] Most publications, however, maintained links with their questionable past by openly advertising practitioners' services and inventions, as well as hours of attendance.

As one might expect, Parker, too, attempted to differentiate himself from the quack practitioner. Not only did he align himself with dental surgeons, such as Tomes, but he claimed much of his work involved repairing the damage inflicted by quacks and reversing the notions that untrained dentists put into the heads of their patients. For example, Parker claimed many quacks were inserting false teeth without first preparing patients' mouths properly. As a result, decaying teeth were not extracted and offered little support to dentures that already sat uncomfortably on swollen gums. In particular, such practices were attacked as they were said frequently to lead the public to believe that all artificial sets were 'useless, uncomfortable, and painful appendages'.[26] Nevertheless, to Parker's advantage, such methods of practice also initially made members of the public less hesitant eventually to approach dentists like him for new sets, as many had come to regard the process as painless, compared with other dental procedures. Those who came to Parker, however, quickly learned that much preparation was required before any replacement teeth could be installed. Many would have been confronted with unexpected and highly undesirable news: a new set of teeth could not be placed adequately on 'loose stumps and gums in soft spongy condition'.[27]

In general, Parker appears to have practised a very aggressive form of dentistry. The preservation of teeth was his first duty, but only as long as they could be made subservient to the uses of mastication and appearance. Alternatively, should teeth have become sufficiently diseased to affect the general health, or so painful as to interrupt the ordinary duties and comfort of life, intervention was necessary. In such circumstances, Parker jumped into action, regarding the 'conservative Dental surgery of the present day [to be] mischievous'.[28] Instead, Parker's patients underwent several weeks or even months of preparation work before artificial teeth could be fitted. Most memorably these included a local clergyman, who came to Parker because an inferior set of artificial teeth made his jaw ache and, equally

[25] J. Robinson, *Dental Anatomy and Surgery* (London, 1846).

[26] Parker, *Remarks upon Artificial Teeth*, 12.

[27] Ibid., 11.

[28] S. A. Parker, 'Contributions to dental surgery', *British Journal of Dental Science* 6, 84 (1863), 268.

important, prevented him from delivering his sermons, which members of his congregation could no longer understand. Like this gentleman, most patients attending Parker's clinics were requested to keep their mouths clean between sometimes dozens of extractions by gargling with an astringent lotion composed of tincture of myrrh and an infusion of roses, before a new vulcanised set was finally manufactured and fitted. While such advice appears to have chiefly benefited the practitioner, those appearing in the greatest pain may have rinsed their mouths with more effective solutions, such as a fomentation of poppyheads.[29]

Though initially working on his own, with no mention of students in the records, Parker, like Dufton at the Ear Hospital and Middlemore at Birmingham's eye charity, aimed to educate local practitioners as well as patients. To begin with, his books requested all those individuals who had been provided with dentures to be patient with their new artificial teeth. In Parker's opinion, most people required some time to get used to a new set. His text also advised patients to place their teeth in a glass of water each night, many having done this only once or twice a week. All were told to rinse their teeth regularly, and, those with prostheses were to remove teeth when cleaning them, as much harm was caused when food was permitted to putrefy and destroy any remaining healthy teeth. In particular, he scolded those patients who damaged their teeth by biting silver coins in order to test if they were genuine. In general, all readers of his text were warned to stay away from quacks or from being led astray by their advertisements.

Though the work of a medical specialist, Parker's book also deals with the secondary diseases caused by rotten teeth. As such, most of his collected cases emphasise that teeth are not detached from the rest of the body, and patients' symptoms often included aching of the face, ear and head, as well as neuralgia, tics, poor digestion and even blindness. Least surprisingly, though rarely commented upon by historians, most of his patients complained of severe lack of sleep, often unable to perform daily tasks, let alone eat regular meals. In contrast with those missing their front teeth, the loss of which clearly affected facial appearance and pronunciation, patients lacking rear teeth were generally less healthy, unable to chew food adequately for the purposes of digestion.

Neither did his patients always appear to be suffering from dental caries. Many often turned up at Parker's dispensary, having already appealed to the town's other medical charities. The most desperate, including many children, were brought with fistulous openings in their cheeks formed when abscesses broke externally, wounds in many cases having discharged pustular matter for months or even years. Others clearly sought help

---

[29] Parker, *Remarks upon Artificial Teeth*, 73.

with fractured teeth following speedy surgery at the hands of numerous unnamed 'quacks' who continued to visit the town in these years. Almost all were said to have been cured by Parker, though in making such claims he comes closest to resembling those whom his published works specifically aimed to condemn. While the stories of suffering recounted in his publications convinced many that the mouth and teeth were in fact more deserving of attention than previously met with, they also suggest why so many dental cases continued to turn up in general hospitals throughout these years. Judging by a steady increase in patient numbers in these years, however, many more than previously appear to have been heading directly to the dental hospital. In time, this alone would lead to the charity's expansion.

By 1871, the institution found itself situated at 9 Broad Street, its name changed to 'The Birmingham Dental Hospital'. As its title denotes, this was a slightly larger institution, treating approximately 3,500 cases annually, and its staff had also expanded. This same year, Charles Sims was appointed a dental officer, his operations undoubtedly being made less painful, though also more dangerous, through the appointment of Lloyd Owen, the hospital's first chloroformist.[30] As before, the hospital staff commenced work at 9 am, only now they received patients every morning of the week, except Sundays. Each month staff treated approximately 350 patients, as many as attended the Liverpool Dental Hospital, though still only a quarter of the number seen at the nation's largest dental hospital in London.[31] Within weeks of the hospital's move to Broad Street, nitrous oxide was also made available to patients, the London dental hospital having introduced the anaesthetic in 1868.[32] While this may have encouraged patients at both hospitals to increase, there is every indication that the Birmingham charity's finances expanded less rapidly, averaging no more than £100 during these years. Nevertheless, appearing to have attained his initial goal, not to mention a large and more remunerative private practice, if not simply a degree of respectability, Samuel Adams Parker retired two years later. In reality, the forty-two-year-old Parker's retirement is more likely to have been connected to ill health and the death of his father two years earlier.[33] The Homeopathic Hospital, with which the dental charity had previously shared premises, incidentally, had also acquired a certain

[30] 'Monthly Meeting of the Birmingham Dental Hospital', *British Journal of Dental Science* 14, 180 (1871), 281–2.

[31] 'Monthly Meeting of the Birmingham Dental Hospital', *British Journal of Dental Science* 15, 191 (1872), 210, 308, 388, 399, 462, 504.

[32] Smith and Cottell, *History of the Royal Dental Hospital*, 24.

[33] Cohen, 'The Birmingham Dental Hospital', 332.

amount of respectability, permitting its governors also to relocate in 1875, in this case, to new premises at Easy Row.

With this move and the adoption of a grander name, the Dental Hospital appeared to mirror developments in the profession more generally. By 1878, with the introduction of the Dental Act, the profession, comprising more than 1,800 dentists nationally (nearly 400 of whom held the Royal College of Surgeon's Licentiate in Dental Surgery), was placed on a far sounder footing.[34] In 1880, further professional gains were made when the school's courses were recognised by the Royal College of Surgeons, pupils being permitted to attend a course of practical instruction at the hospital upon payment of 5 guineas. The first to do so was F. W. Richards, who came to the hospital in 1883 and eventually became the school's dean in 1896, a post he held until 1911.[35] Four more pupils came in 1885. The following year a student dental society was even formed. Over the next fifteen years another sixty students attended the hospital's dental school to obtain their degrees. While most came from the Midlands and twenty-five alone were Birmingham residents, fourteen came to Birmingham from outside the region, six of these from outside England, including South Africa. Clearly attracting both pupils and patients, staff frequently left most of the latter in the hands of students working under the supervision of one of three dental surgeons. All dental surgeons, on the other hand, were now required by hospital regulations to be registered under the Dental Act of 1878, practised dentistry exclusively and never advertised.

By this time, the management of the hospital had been placed under the guidance of Charles Sims, whose main concern was to increase the hospital's finances. In 1879, for example, though patients approached 6,000 in number, income was still only £187, £51 (27 per cent) of this amount comprising subscriptions, most of the remaining costs being covered by a timely legacy.[36] Unlike other medical charities, the hospital still appeared to be relying to an unusual extent on the small amounts patients deposited in the institution's 'poor box'; during this particular year, hospital visitors and grateful patients left random donations totalling £7, which, though no different from amounts collected at other institutions, comprised a more substantial percentage of income (4 per cent) than at other voluntary hospitals. In general, the institution's governors blamed this poor performance on 'the multiplication of small charities' in

34 E. Smith, 'The Royal Dental Hospital of London: the first fifty years', *Dental Historian* 33 (1998), 27.

35 UBSC, Annual Report of the Birmingham Dental Hospital, 1910.

36 UBSC, Annual Report of the Birmingham Dental Hospital, 1879, archive collection.

Birmingham.[37] Moreover, the answer did not lie in initiatives such as Hospital Saturday, which brought 'small charities' like the dental hospital less than was collected in its charity boxes during the 1870s. As a result, the hospital's administrators continued to rely exclusively on its eighty subscribers, including local men (57), women (4), businesses (18) and an orphanage. The inclusion of the names of several particularly influential local families, such as the Ansells, Cadburys, Chamberlains, Elkingtons and Gillotts, however, provided governors and staff with considerable hope for the future.

Largely due to this influential list of subscribers, the hospital was reconstructed in 1882 at 71 Newhall Street, though its balance hardly surpassed £50 in 1879.[38] Parker and Sims, along with Thomas English, continued to serve as consulting surgeons to the institution which, by 1888, had a staff comprising four honorary dental surgeons, four assistant surgeons (including John Humphreys, who joined the staff in 1883)[39], seven anaesthetists, a museum curator and a demonstrator. In that same year, staff treated 9,864 cases and performed 8,783 extractions, still only 464 procedures being performed with the help of anaesthetics. The surgical committee was particularly pleased with an increase in conservative treatment, the hospital having provided 1,169 fillings and another 1,992 patients primarily with advice. The governors would have been as pleased that all of this work cost only £383, which was matched by available income. While subscriptions continued to raise a disappointing £100, the hospital's finances promised to keep up with expenditure from now on as its management committee had decided to introduce a 6d registration fee for all patients above ten years of age. In 1888 alone this raised £87 and would lead registration fees to surpass subscriptions in importance in subsequent years. The privileges of 124 subscribers, on the other hand, had been reduced, each guinea subscription entitling donors to only six admission tickets instead of eight.

Not surprisingly, over the next decade subscriptions hardly increased, reaching £153 in 1897. All other figures, however, continued their steady rise. Income was £1,028, half of which came from registration fees. With four additional dental surgeons, the annual number of operations easily surpassed 20,000 and staff and students were attempting and learning new procedures, including minor plastic surgery, as well as crown and bridge work. In general, the staff's workload continued to increase, though

[37] 'Hospital reports and case-book', *British Journal of Dental Science* 19, 238 (1876), 178.

[38] UBSC, Birmingham Dental Hospital, Surgical Committee, Minutes 1880–7, MSS II (formerly rq RK3 B5).

[39] A. W. Wellings, 'John Humphreys, of Birmingham: an appreciation', *British Dental Journal* 63, 1 (1937), 49–50.

noticeably fewer patients attended for advice alone. Perhaps for this reason, despite the institution's good work, the following year a mechanical department was opened, and the management committee was convinced that 'the supplying of artificial teeth will form one of the most important parts of the future work of the Hospital'.[40]

While this may seem to suggest that dental cases over the last quarter of the nineteenth century progressively concentrated at the Dental Hospital, a large number of dental cases continued to be treated at the town's other hospitals throughout this period. For example, in 1875, staff at the dental department at the Children's Hospital treated fifty-two patients, compared with twenty-one the previous year.[41] The following year this increased to ninety-two cases. Though annual reports after 1884 no longer record exact numbers of dental cases, they state that both the eye and teeth departments were regularly consulted by other members of the hospital's medical staff, not to mention students. For example, in 1892, the Children's Hospital received a letter from Mr Fowler asking permission for the dental students at Queen's College to attend all operations of the mouth undertaken at the hospital. In future, the resident medical officer was requested to forward a list of these operations to the dean.[42]

Though similar work was also being carried out by staff at the General Hospital, dental treatment at the town's largest hospital only really commenced in 1881, the year the Dental Hospital, as opposed to a dispensary, was founded in Birmingham. At that time, the General's medical committee determined the hospital's dental surgeon was to be a Licentiate of the Royal College of Surgeons and to attend one hour each day to see all tooth cases.[43] Despite a decision in 1883 to send all such patients to the Dental Hospital when possible, their numbers never entirely disappeared from annual reports in these years. In 1884, for example, surgeons there still extracted 1,000 teeth, though this had noticeably declined from 3,000 a year earlier and nearly 4,000 for each year since 1870.[44] Nevertheless, in the early 1890s dental cases were again increasing, reaching 3,329 in 1892. All dental patients were to be seen before noon and, at all other times, only accident cases were treated at the hospital. Presumably, the General's dental cases had always belonged to the latter category. However, special equipment was also added to the hospital's dental department. For example, a

[40] UBSC, Birmingham Dental Hospital, Surgical Committee, Minutes 1898–1903, MSS 14.

[41] BCLA, Children's Hospital, Annual Report, 1875, HC/BCH/1/14/3.

[42] BCLA, Medical Committee Minutes, 1877–93, HC/BCH/1/4/3.

[43] BCLA, General Hospital, Medical Committee Minutes, 1876–84, HC/GHB/70.

[44] BCLA, General Hospital, Annual Report, 1884, GHB 429.

small motor sufficient to work a dental engine and an attachment for cautery purposes was recommended to be placed in the hospital's large operating theatre. Interestingly, only in 1898 was the dental surgeon requested to examine the teeth of nurses, as well as those of other hospital servants.

Despite the amount of dental work carried out elsewhere, the Dental Hospital was recognised as the premier dental institution of not just Birmingham, but the whole of the Midlands. Though students were observing dental cases at other hospitals, in the 1890s, the Dental Hospital was considered reputable enough to gain affiliation with the town's medical school. As a result, when the University of Birmingham was formed in 1900, it also became the first institution of higher education in the country to grant a degree in dental surgery. More generally, it is thought to be one of the first dental hospitals in the English-speaking world, being preceded only by a similar charity in Baltimore, Maryland, and the London Institution for Diseases of the Teeth, which were founded in 1839 and 1840 respectively.[45]

Although the latter institutions are mentioned in other histories of dentistry, rarely do these same studies mention Birmingham's earliest dentists. While a decision in 1884 to confer a knighthood on Edwin Saunders, one of the founders of the London Institution for Diseases of the Teeth and the first member of the profession to be so distinguished, was another indication that the status of dental practitioners had changed significantly in only a few decades,[46] similar awards were not bestowed on Parker and Sims. Having remained a consulting dental surgeon to the hospital until the early 1890s, Parker died in relative poverty and obscurity in 1896, overdosing on laudanum, aged only sixty-two.[47] Sims's reputation, on the other hand, was longer lasting. Given that he was the chief surgeon at the Dental Hospital at the time it affiliated with Birmingham's medical school, he retired in 1892 after serving in the capacity of Professor of Dental Surgery at Queen's College. Though he never published anything significant in his field, he was officially recognised by his peers as one of the founders of the British Dental Association and served as President of the organisation's central counties branch. An obituary published in the *Dental Record* describes him as the 'foremost dental surgeon in Birmingham thirty years ago'.[48] He remained a consultant at the Dental Hospital until the first years of the twentieth century and died of a stroke in Finchley Nursing Home in 1906.

[45] Smith and Cottell, *History of the Royal Dental Hospital*, 13–14.
[46] Smith, 'The Royal Dental Hospital of London', 25.
[47] Spencer, 'Notes on the history of dental dispensaries', 52.
[48] *Dental Record*, 1 October 1906.

A MORAL INSTITUTION:
THE SKIN HOSPITAL

While the Birmingham Dental Hospital was one of the first hospitals of its kind in England, in some respects, the town's skin, or lock, hospital was a comparatively late development. Nationally, lock hospitals had been among the earliest of specialist institutions, skin and venereal ailments having been grouped together from a very early period. With a rise in the incidence of syphilis in the sixteenth century, afflictions of the skin became inextricably linked with those of venereal disease.[49] The establishment of a Lock Hospital for the treatment and cure of venereal disease met an obvious need with the widespread and growing incidence of such disorders in eighteenth-century London.[50] Of equal importance to the development of dermatology was the foundation of fever hospitals, as appeared in London, for example, in 1802. Less researched has been the connection between skin diseases and Poor Law institutions, where many of these conditions would have first been noticed.

Nevertheless, until the eighteenth century, an individual's skin was looked upon merely as a mirror that reflected the state of internal disease, rather than as an important organ, subject to its own disorders.[51] Pain or itching in the dermis was routinely treated, and occasionally mitigated, with generous applications of fresh butter and rose water.[52] Those skin cases that were linked to venereal infections were treated with more hazardous remedies; mercurial inunctions, or ointments, were the grand panacea for all syphilitic infections.[53] Those writing on the subject in the mid- to late nineteenth century, dated the modern treatment of such ailments to John Hunter's work of 1786, which argued for the existence of specific 'morbid animal poison' or 'virus', though the Scottish surgeon continued to regard syphilis and gonorrhoea as varieties of the same disease.[54] Despite such developments, with sexual vice weighing heavily on the public mind during the French revolutionary wars, the physical cure of venereal disease

[49] K. Siena, *Venereal Disease, Hospitals and the Urban Poor: London's 'Foul Wards,' 1600–1800* (Rochdale, NY, 2004), 19.

[50] D. Andrew, 'Two medical charities in eighteenth-century London: the Lock Hospital and the Lying-in Charity for Married Women', in *Medicine and Charity before the Welfare State*, ed. J. Barry and C. Jones (London, 1991), 89; Siena, *Venereal Disease, Hospitals and the Urban Poor*.

[51] M. Morris, 'An address delivered at the opening of the Section of Dermatology', *British Medical Journal* (18 Sept 1897), 697.

[52] D. Turner, *A Treatise of Diseases Incident to the Skin* (London, 1736), 67.

[53] Siena, *Venereal Disease, Hospitals and the Urban Poor*, 22–4.

[54] L. Parker, *The Modern Treatment of Syphilitic Disease* (London, 1859), 1.

was not enough. With gainful employment for women on the wane, and many more turning to prostitution, the moral reformation of patients had become equally important.[55] Significantly, a year after Hunter's work appeared, the Lock Hospital in London established a sister institution, the Lock Asylum, with the aim of containing such cases and promoting their moral reform.[56]

Treatment of all other skin afflictions, on the other hand, eventually changed with the appearance of Robert Willan's *On Cutaneous Diseases* in 1808. By the time Willan's work appeared, the Edinburgh-educated physician was already recognised as an authority on the skin, the surgeon having been awarded the Fothergillian gold medal in 1790 by the Medical Society of London for his outline of the arrangement and description of skin diseases. With his later work came the first proper definitions of skin disease and attempts to describe their general divisions and treatment, many of his observations having been made amongst the malnourished inmates of the Public Dispensary in Carey Street and the London Fever Hospital, where he held appointments. After his death, Willan was succeeded at the Dispensary by Thomas Bateman, who subsequently helped complete his former master's work.[57] In turn, the Carey Street Dispensary acquired a reputation for skin diseases and gradually began to attract qualified practitioners in search of postgraduate training in dermatology. One of those periodically attending the charity was Anthony Todd Thomson, who went on to found the Chelsea, Brompton and Belgravia Dispensary in 1812. In successive years, conditions of the skin observed at these and other institutions underwent ever more detailed classification, climaxing with Erasmus Wilson's *Diseases of the Skin* (1851), which captures the sort of picturesque and ever-changing nomenclatures for which he and other English surgeons were becoming famous.[58] Consequently, despite founding the *Journal of Cutaneous Medicine and Diseases of the Skin* in 1867, Wilson left nineteenth-century dermatology, not to mention his students at the Middlesex Hospital Medical School, in a confused state.[59]

While the English contribution to dermatology at this time was essentially clinical, using classification in order to aid diagnosis, the French systematised this knowledge. Additionally, the French school revived the

---

55 Siena, *Venereal Disease, Hospitals and the Urban Poor*, 182.

56 Andrew, 'Two medical charities in eighteenth-century London', 93–4.

57 T. Bateman, *A Practical Synopsis of Cutaneous Diseases, According to the Arrangement of Dr Willan* (London, 1836).

58 R. M. Hadley, 'The life and work of Sir William James Erasmus Wilson, 1809–84', *Medical History* 3 (1959), 215–47.

59 Morris, 'An address delivered at the opening of the Section of Dermatology', 698.

idea that many venereal afflictions were the result of 'ordinary irritation'.[60] Both schools of thought generally observed patients' symptoms more closely than had previously been the case. The German approach by the mid-nineteenth century, on the other hand, was pathological. By redirecting attention to the processes of skin disease, German practitioners simplified dermatology; many conditions which had formerly been looked upon as distinct affections were finally recognised as little more than different stages of the same disease. Furthermore, medical practitioners learned to distinguish between the primary lesions of a disease and those that resulted from various secondary causes, whether attributed to scratching or the invasion of parasites, as in the case of scabies and ringworm. Nevertheless, many dermatological afflictions remained closely tied to venereal disease, syphilis, most famously, often mimicking other diseases of the skin. So, too, in these years, was tuberculosis proven to be accountable for *lupus vulgaris*, as well as a number of other skin afflictions, thereby further reducing the mystery formally associated with the field of dermatology. Not surprisingly, at the same time, dermatology ceased to be an 'error to students and an affliction to practitioners'.[61]

Despite being a relatively late development locally, the skin hospital in Birmingham commenced on a very small scale around the time that skin specialists had been gaining recognition and respectability in England. Between 1819 and 1899, at least thirty such institutions were founded in Britain.[62] London alone 'had broken out in a rash of skin hospitals', possessing eleven such institutions;[63] in some cases this led practitioners formerly in favour of specialisation to describe it as 'specialism rampant', or an 'evil excess'.[64] Among these was the first skin hospital in Britain, the London and Westminster Infirmary for the Treatment of Diseases of the Skin (1819), as well as the most influential, Blackfriars Hospital, founded in 1841 by James Startin, who also happened to be born in Moseley, Birmingham, and worked at the General Hospital, first as a surgical dresser to Joseph Hodgson, then two years as House Surgeon;[65] a key figure in the history of dermatology, Startin is presumed to have developed the first

[60] Parker, *Modern Treatment of Syphilitic Diseases*, 2.

[61] Morris, 'An address delivered at the opening of the Section of Dermatology', 700.

[62] A. Rook, 'Dermatology in Britain in the late nineteenth century', *British Journal of Dermatology* 100, 1 (1979), 4.

[63] B. Russell, *St John's Hospital for Diseases of the Skin, 1863–1963* (London, 1963), 19.

[64] E. Mackey, 'Remarks upon the special study of skin diseases; with an analysis of three hundred cases', *Birmingham Medical Review* 4 (1875), 241.

[65] A. Rook, 'James Startin, Jonathan Hutchinson and the Blackfriars Skin Hospital', *British Journal of Dermatology* 99 (1978), 216.

ether inhaler to have been mass produced commercially.[66] Equally important to the development of the field of dermatology was St John's Hospital
for Disease of the Skin, founded in Leicester Square in 1863 by John Laws
Milton. Its history, however, was a very 'stormy' one, as the subject still
required 'delicate handling', and 'delicate' hardly described Milton's usual
mode of treatment.[67]

Though a skin institution in Birmingham did not emerge until the
1880s, other local medical practitioners working decades earlier helped
develop the ideas and methods that would be implemented in the curative regime of such a hospital. One of these individuals was Langston
Parker, father of the innovative dentist, Samuel Adams Parker. During his
long career at the Queen's Hospital in Birmingham, Langston Parker had
devoted years and considerable attention to venereal diseases. By the time
his work on the treatment of syphilitic diseases was published in 1859, he
had devoted nearly twenty years to the therapeutics of syphilis, treating
approximately 8,000 cases, many of which are discussed in the pages of his
printed works. In general, the system of treatment Parker advanced was
less reliant on mercury, though he continued to regard the element as the
most powerful therapeutic agent against syphilis. Instead, he advocated
the daily cleaning of venereal sores with a soft sponge and tepid water.
Primary syphilitic sores, in his opinion, were best treated with strong aqueous solutions of opium, weaker mixtures of silver nitrate, copper nitrate
or even port wine.[68] Unlike Hunter, Parker did not regard gonorrhoea as
a variety of syphilis, but attributed it instead to a variety of causes, ranging from sexual intercourse to cayenne pepper.[69] More influential was his
belief that syphilitic ulcers healed more quickly when a patient was put
on a low diet and into bed. If lucky enough to gain admission to a medical institution in the two decades after the appearance of Parker's treatise,
many of Birmingham's venereal patients continued to enter the town's
general hospitals, rather than a specialist skin hospital, though the majority of skin cases, as in London, surely attended local dispensaries and the
workhouse.

Much like the Midlands' other specialist hospitals, the Birmingham
Skin Hospital first opened as a dispensary at Newhall Street in 1881. In
that year, medical staff, comprising two consulting surgeons, two physicians and an ophthalmic surgeon, attended four days a week and treated

[66] E. T. Mathews, 'Startin's pneumatic inhaler', *Proceedings of the History of Anaesthesia Society* 32 (2003), 20–6.

[67] Russell, *St John's Hospital*, 23.

[68] Parker, *Modern Treatment of Syphilitic Disease*, 10.

[69] Ibid., 56.

270 patients annually.[70] In its second year, the hospital's patients rose to 968 and accounted for nearly 5,000 attendances. The majority were lock, or venereal, cases, though skin cases were increasing, the most common cutaneous afflictions being eczema (195), psoriasis (53) and scabies (43), and were treated with regular vapour baths. Of the venereal cases, gonorrhoea was by far the most numerous, with the hospital treating 177 registered sufferers. Regarded as the least respectable of individuals, venereal patients found a staunch supporter in Thomas Heslop, who championed the charity, and, as a result, simultaneously ranks as Birmingham's most prolific hospital-builder.

Unfortunately, though Heslop occupied the post of consulting physician to the charity for four years, he did not witness the completion of the charity's purpose-built premises on John Bright Street in 1886, having died the previous summer while on holiday in Scotland.[71] Nevertheless, Heslop had clearly timed the establishment of the hospital well, its doors having opened the same year the Dermatological Society of London, the first British dermatological society, was founded. Though the membership of this professional body never exceeded sixty, the Birmingham skin charity's staff treated many times this number of patients each week, admitting approximately 2,000 new cases annually.

Soon after its new premises were completed in 1886, staff at the institution treated 2,000 patients, who suffered from both venereal diseases and various skin complaints, many of the latter being the direct result of working conditions. By 1889, though the charity still maintained only a dozen beds, its staff managed to attend nearly 4,000 patients annually, treatment involving primarily the use of medicinal baths and a generous course of largely ineffectual, if not dangerous, pharmaceutical remedies. At the time, staff at St John's Hospital for Diseases of the Skin in London treated similar numbers with nearly three times the number of beds.[72]

Despite a noticeable growth in the number of venereal cases in the region, there were other reasons for establishing such a hospital in Birmingham. To begin with, few such cases were treated at other local medical charities, including the Women's Hospital. Venereal disease was also seen as a threat to children, and the charity was originally justified as an attempt to ensure the health of a future generation, and not necessarily their infected parents. As already mentioned, the effects of industry equally heightened the need for a skin hospital locally. During its earliest years, half of the skin cases treated at the hospital were regarded as industrial injuries, due to

---

[70] UBSC, Skin Hospital, Annual Report, 1881.

[71] *Oxford Dictionary of National Biography*, accessed online 25 July 2008.

[72] Russell, *St John's Hospital*, 59.

both the processes and products used in factories and workshops. Among those workers appearing most often at the hospital were brass-workers, French-polishers, jewellers and domestic servants. In fact, venereal disease was becoming less common at the institution during the two decades following its foundation. By 1896, 75 per cent of all skin cases treated at the hospital were industrially related.[73] As a result, in this year the institution's name was changed to the 'Hospital for skin and urinary diseases', a move, without a doubt, calculated to elevate the prestige of the institution.

The reputation of the Skin Hospital, like that of the Dental Hospital, was only further elevated when the charity was designated a teaching hospital in 1896. Dispensing 35,000 prescriptions a year to 6,000 patients, the Skin Hospital offered much clinical material to pupils at the medical school. Of additional interest is that many of these prescriptions, as at the Women's and Children's hospitals, were prepared by female dispensers before the end of the nineteenth century. In contrast, despite popular representations of venereal sufferers, male patients outnumbered females by more than four to one, while local women were nearly twice as likely to seek treatment for skin complaints at the charity. In general, whether seeking treatment for skin or dental conditions in the last years of the nineteenth century, many more of Birmingham's inhabitants headed to qualified specialist practitioners and their associated institutions, the reputations of which had witnessed a dramatic transformation in less than half a century.

[73] BCLA, Skin Hospital, Medical Committee Minutes, 1890–1928, MS 1918.

CHAPTER 6

Late-Nineteenth-Century Reorganisation: The Associated Teaching Hospitals

B EDS AND PATIENT NUMBERS were set to increase at all of the Birmingham voluntary hospitals, not just the emerging specialist institutions, in the last decades of the nineteenth century. As a result, much hospital reconstruction would characterise the 1880s and 1890s. Starting with the appearance of the last specialist hospital to be founded in Birmingham in the nineteenth century, this era witnessed the construction of purpose-built facilities for both the Skin and Ear hospitals. By the last decades of the nineteenth century, as has been argued elsewhere, the battle for the acceptance of specialisation had been fought and won throughout most of continental Europe.[1] A similar victory would soon be claimed by specialists in Britain. Far greater transformations, however, were in store. Not only would ever-greater numbers of Birmingham's inhabitants experience hospital treatment during these years, but many more would become familiar with the escalating costs associated with hospital medicine. For many of the town's inhabitants, health care became more and more closely associated with hospital medicine.

LOWER OPERATING COSTS:
THE ORTHOPAEDIC HOSPITAL

By the early 1880s, the Orthopaedic Hospital continued its steady growth and was still the only hospital of its kind outside London. In 1877, the charity had moved into a building vacated by the Institution of Mechanical Engineers which, like many of Birmingham's most eminent doctors, including John Ash, William Bowman and Joseph Hodgson, had abandoned Birmingham for the capital. The new hospital that opened in its place initially had twelve beds, with room for eight more. However, given that most individuals at the institution were children, occasionally twice that number resided at the hospital, staff having frequently allocated two young patients to each bed during this period.[2] The work of the hospital was largely in the hands of Miss Dorrell, the matron, and a dispenser, Mr Rogers, and a member of the Freer family was still to be found on its staff. While Thomas Freer together with Bell Fletcher, another of William

---

[1] Weisz, *Divide and Conquer*, xxi.
[2] BCLA, Orthopaedic Hospital, Annual Report, 1877–8, HC/RO/Box 14.

Sands Cox's first students, served as honorary surgeons, the appointment of Thomas Heslop to the consulting staff in 1879 considerably improved links between the Orthopaedic and Children's hospitals.

Like the Children's Hospital a decade earlier, the work of the orthopaedic institution increased significantly in these years. Staff saw 6,156 patients in 1879, 2,054 being new cases, another 372 being admitted as inpatients. Of the new cases, 1,539 (75 per cent) were said to have been cured in the year, another 342 (17 per cent) greatly relieved, while 173 continued their treatment beyond a year.[3] In general, early treatment was advocated to avoid permanent disability, or unemployment, later in life. While patients were said to come from all over the region, deformities from the mining districts around Birmingham were said to 'abound to a most lamentable extent'.[4] In total, 636 (31 per cent) patients came from neighbouring towns, most residing in Kidderminster and Wolverhampton, though few subscribers from these areas are recorded in subscription lists. The principal afflictions recorded in the annual reports for 1879 are spinal cases (269 or 4.3 per cent of those undergoing treatment in this year), bowed legs (424; 6.9 per cent), club feet (252), contracted heels (443), flat feet (195), hernia (105), knock knees (414), paralysis (138), rickets (296), weak ankles (292) and hip disease (64). For at least 269 of these cases, treatment had improved with the recent introduction of Plaster of Paris jackets to correct curvature of the spine, and hospital surgeons had also begun to divide bones surgically in cases of knock knees and bowed legs.

In general surgical cures became more common in the 1830s. By this time, practitioners with national reputations such as James Syme, the leading surgical consultant in Scotland, were advocating excision in place of amputation.[5] His *Treatise on the excision of diseased joints* (1831) refers to fourteen elbow resections at a time when amputation was regarded as the only safe and efficient 'means for removing diseased joints which did not admit of recovery'.[6] Taking much longer than amputation, and far more painful, excision became a realistic option for many patients after the development of anaesthesia. More influential than Syme's work was that of William John Little, whose own paralytic club-foot was cured as a result of a tenotomy performed by the young German surgeon Georg Stromeyer in Hanover in 1830.[7] Encouraged by Stromeyer, Little popularised the subcutaneous

3 BCLA, Orthopaedic Hospital, Annual Report, 1879–80.

4 Ibid.

5 J. Syme, *Treatise on the Excision of Diseased Joints* (London, 1831).

6 LeVay, *History of Orthopaedics*, 99.

7 Ibid., 500–1; P. Stanley, *For Fear of Pain: British Surgery, 1790–1850* (Rodopi, 2003), 246.

procedure from Bloomsbury Square, London, at his 'Infirmary for the Cure of Club Foot and other Contractures'; founded in 1838, Little's infirmary became the Royal Orthopaedic Hospital in 1845, moving to Hanover Square in 1855.[8] Orthopaedics entered a third, more adventurous, phase with the work of Lister, who has also been saddled with responsibility for some of the excesses that followed.[9] Nevertheless, while surgeons often promised disabled individuals a quick fix, treatment for many patients of orthopaedic charities would continue to involve repeated surgical interventions over many years well into the next century.[10]

By 1879, staff at the Birmingham Orthopaedic Hospital performed 965 operations. Additionally, they distributed 606 instruments, costing £400, to the charity's poorest patients. As these developments perhaps suggest, surgical instruments continued to be one of the charity's main expenses, though a greater propensity to operate promised to reduce these costs as the knowledge and confidence of medical staff increased. Unlike in previous decades, however, increases in expenditure no longer caused the hospital's governors considerable anxiety, finances having improved substantially since the institution's move to Newhall Street. Annual income now surpassed £1,000, thanks to increasing contributions from both the Hospital Saturday and Sunday organisations, which raised more than £561 of income.[11] Subscriptions still generated approximately a third of funds, though the total realised from these regular payments had now been surpassed by income from investments. Significant sums also came directly from patients, who now paid for services, as did the committee of the Ear Hospital, who had become the institution's subtenants in 1877. The orthopaedic charity would continue to collect £100 annually from their medical tenants until 1883.

Though income would suffer due to a slowdown in the local economy over the next years, if anything, the work of the hospital only increased with the occurrence of such trade depressions. Consequently, the hospital's debt would climb in this period, as would its death rate, due almost entirely to the propensity of staff to undertake new and riskier surgical procedures that promised to cure a number of formerly chronic conditions and save many younger patients long and uncomfortable months in braces or splints. Furthermore, besides appointing an anaesthetist in 1885, administrators also constructed the hospital's first mortuary.[12] The following year, beds

[8] LeVay, *History of Orthopaedics*, 125–6.

[9] Ibid., 107.

[10] White, *Years of Caring*, 196–7; Borsay, *Disability and Social Policy*, 55.

[11] BCLLS, Orthopaedic Hospital, Annual Report, 1879.

[12] BCLLS, Orthopaedic Hospital, Annual Report, 1885.

had increased to eighteen. Three years later, in 1889, when Birmingham achieved city status, the hospital gained two additional wards and an operating theatre, before it was also granted its 'Royal' title.[13] With the development and perfection of antiseptic methods and the gradual introduction of aseptic operating techniques, hospital staff began to perform many more successful operations on an increasing proportion of adult patients. Surprisingly, mortality figures did not increase in line with the charity's status, though subscriptions predictably did.

So, too, did the number of students on hospital wards begin to climb. Though private pupils of the hospital's own surgeons had been admitted to the institution since 1877, a greater student presence was promised when the hospital became affiliated with the medical school in 1892. While only a few senior medical students were appointed clinical assistants to the institution for a period of three months, many others came to the hospital in order to attend operations and observe cases in its outpatient clinic. The following year, with the appointment of a masseuse, nursing staff commenced their own training courses in massage. Soon after, Miss Wolseley, the dispenser, was permitted to take her own pupils. Members of the medical staff, on the other hand, were provided with an opportunity to learn about the 'new photography' at first hand when radiography equipment was acquired in 1896, less than a year after Wilhelm Roentgen publicised his revolutionary work on x-rays.[14] From this date, staff more easily agreed on the treatment required by their most difficult cases. In the last years of the nineteenth century, the hospital admitted 1,500 new cases annually, nearly half of which were inpatients. In order to facilitate the greater accommodation of all cases, governors made the decision to move the outpatient department to separate premises at 22 Great Charles Street in 1898.

NURSING INNOVATIONS:
THE QUEEN'S HOSPITAL

With 130 beds in 1875, the Queen's Hospital also undertook a remarkable amount of work. Besides treating more than 2,000 inpatients yearly, it dispensed to and dressed the wounds of an additional 25,000 outpatients. Staff had also begun to undertake a considerable amount of work at specialist clinics held once a week, treating obstetric (237), midwifery (334), dental (742) and ophthalmic (442) cases.[15] Dental cases were the most numerous patients at these early clinics, though they declined the following year, while maternity, obstetric and ophthalmic cases increased

---

[13] White, *Years of Caring*, 29.

[14] BCLA, Orthopaedic Hospital, Annual Report, 1896, HC/RO/Box 14.

[15] BCLLS, Queen's Hospital, Annual Report, 1875.

more rapidly. Attending the hospital's outpatient department on clinic days, some of these specialist cases were subsequently admitted as inpatients. For example, fifty-seven of 360 ophthalmic cases gained admission to the hospital in 1876, staying on average fifteen days. On such occasions, many became teaching subjects. Joseph Priestley Smith, one of Birmingham's best-known ophthalmic surgeons, also began to use such cases in his evening clinical lectures, as well as the courses in ophthalmic surgery he delivered to senior students and resident officers shortly after he was appointed to run the hospital's ophthalmic clinic in 1874. Late to enter medical school, though having attended both Sydenham and Queen's colleges, Priestley Smith first served an apprenticeship in mechanical engineering, skills which he relied on to invent instruments useful to his optical research, including a perimeter and a tonometer, the latter of which was used to measure intraocular pressure;[16] an experimental model of a practice eye made by R. Bailey of Birmingham (c. 1900), was similarly based on his instructions.[17] By 1886, Smith was treating nearly 2,000 eye cases annually at the hospital.[18] Undoubtedly, these included a number of the cases, the details of which were to reappear in his well-circulated studies on glaucoma, a subject on which he was subsequently invited to lecture at the eighth and ninth International Congress of Ophthalmology in 1882 and 1886 respectively.

With an increase in all teaching cases at the hospital, from this period onwards there developed considerable pressure to rebuild and innovate, resulting not only from the institution's centrality to patient care in Birmingham, but also the charity's proven role in educating a new generation of local practitioners. As a consequence, a Jubilee appeal was made in 1886 in order to place the hospital 'in accord with the requirements of modern science'.[19] Despite these plans, further additions to the hospital in these years were limited. With the appointment of Dr George Crooke in 1889, an attempt was made to use the pathological department to better effect, but the majority of Crooke's time was spent teaching medical students. The operating theatres were also improved, with two new anterooms being added in 1894, one for anaesthetising patients, the other for recovery. An additional £150 was also spent on new instruments in 1897. The same year, George Myers presented the hospital with a bacteriological

[16] J. Priestley Smith, *On the Pathology and Treatment of Glaucoma* (London, 1891), 10.

[17] A model of this eye is included in the museum of the British Optical Association, the College of Optometrists, London. I am grateful to Neil Handley for drawing my attention to this.

[18] BCLLS, Queen's Hospital, Annual Report, 1886.

[19] Ibid.

microscope valued at 20 guineas. A photographic darkroom was set up a year later and a Roentgen apparatus was finally acquired in 1899. Modern science, as a result, made its appearance in these years, but more gradually than staff had hoped.

More important improvements relied less on advances in science than in technology. For example, the installation of a telephone considerably improved communications at the institution. Sanitary conditions also improved in 1893 when the bathrooms were moved outside of wards and new bedsteads installed. Four years later all beds were provided with locally produced, wire-woven mattresses. Other alterations included an extension of the two rear wings of the hospital's main building in 1889. Although set up in 1874, the ophthalmic department was also enlarged in 1892 and a proper water supply was installed the following year. By 1895, the governors claimed the hospital had never before been in as efficient a situation.[20] Electric lighting was installed three years later and, by 1899, electricity serviced the entire building, as well as the operating theatres where it was used to facilitate cauterisation and drilling.

Far more important to the health and comfort of staff was the erection of a separate nurses' home, with room for twenty-six nurses and a caretaker, completed in 1889. As a result, the institution's domestic servants were no longer housed underground, and an asphalt tennis court encouraged all residents to spend greater time outdoors. Despite such amenities, many nurses continued to leave each year because of the noticeably slow rate of promotion at the institution. In general, the hospital was unable to offer all of these women positions commensurate with their abilities. To rectify this situation, a private nursing institution was established in association with the hospital in June 1891. The department opened in October and commenced work with six nurses who had each completed two years of preliminary training. By the end of its first year, the department employed eleven nurses who had attended 150 external cases since its foundation only months earlier.[21]

Probationers joining the hospital now signed up to a four-year training agreement. In their first year, nurses learned primarily in practical classes. In their second, they attended lectures in medicine, surgery, obstetrics and ophthalmics which were delivered by members of the hospital's senior medical staff.[22] Lectures originally commenced in 1881, and a nurses' library was begun in 1893 as the result of a £100 donation. By 1897, the first seven nurses were instructed in massage. During their periods of training, nurses

[20] BCLLS, Queen's Hospital, Annual Report, 1895.

[21] BCLLS, Queen's Hospital, Annual Report, 1891.

[22] BCLLS, Queen's Hospital, Annual Report, 1899.

did not receive a salary, only a uniform. All were to complete two years on the hospital's wards, attend lectures and pass an examination before joining the nursing department. At the end of training, nurses ceased to have hospital responsibilities. In addition to a salary, all nurses also received a share of the Bonus Fund, which comprised registration fees and 10 per cent of the department's gross profits, which was divided in proportion to the time each member annually spent in the department. In return for the use of hospital facilities, the external nursing department also contributed £90 to its funds each year. More important than this, however, was the new department's contribution to the standard of nursing at the institution. By 1893, the Queen's Hospital was employing thirty nurses, or one per 4.5 beds. Ten years earlier there were only twenty-one, or, with 120 beds, one nurse per 5.75 beds. Despite these improved figures, the nursing day did not become any easier. In general, nurses worked from 6 am until the commencement of evening service in the chapel at 9:30 pm. Most enjoyed only two hours of free time daily which were spent either in exercise, rest or chat. Not surprisingly, the nurse's weekly half-holiday remained very precious, though conditions at the well-funded Queen's Hospital were better than at many smaller specialist institutions.

LECTURING LOCALS:
THE MEDICAL SCHOOL

Medical education itself further transformed and consolidated in these years. Surprisingly, however, while record-keeping at the Queen's Hospital improved with the departure of Sands Cox, and the five-year tenure of Henry Burdett (1868–73), the records of the medical school remained poor until it merged in 1892 with Mason's College, a liberal arts college founded by the steel-pen manufacturer and philanthropist, Josiah Mason, twelve years earlier. As a result, the only records relating to the school prior to this date relate to student enrolment and lectures held at the institution.

As one would expect after several turbulent years, student numbers improved after Birmingham's two medical schools merged. Each decade also saw enrolment hit new levels. For example, from 1876 to 1878 new entrants doubled from fourteen to thirty. The majority of these young men, ranging from seventeen to thirty years in age, came from neighbouring Midlands counties, only one coming from as far away as Oxford.[23] In 1881, their numbers peaked at thirty-nine before dropping to a low of twelve in 1884, the same year the college also attracted its first foreign student, Herman Vermaak, a South African, who had passed his preliminary examinations at the Cape of Good Hope University before travelling to

---

[23] UBSC, Queen's College, Medical Department, Students' Register, 1876–92.

Birmingham to continue his studies. For the remainder of the decade, student numbers again climbed, reaching forty-nine in 1889. Though students from outside the Midlands now tended to come from further afield, including Devon and York, proportionately their numbers declined, the majority of young men enrolled at the school coming from Warwickshire and adjoining counties.[24] For the remainder of the nineteenth century, the school attracted on average forty new students annually, 10 per cent of whom were the sons of physicians.[25]

Despite students' middle-class backgrounds, discipline is reputed to have improved only with the appointment of a benevolent dictator, Bertrand Windle, to the post of Dean in 1891.[26] More than likely, Windle's efforts to update student amenities, including the dissecting rooms, library and common room, described as merely 'a shelter from the rain and wind', certainly helped improve morale in these years.[27] Despite alleged improvements in student behaviour, the ages and provenance of students does not appear to have changed considerably in these years.[28] Their academic performance, however, did. Nationally, medical students at Mason's College had a rejection rate that was lower in seven of nine medical examinations. The fewest rejections (3 per cent) were recorded in first-year elementary physiology, a fact that would have pleased John Berry Haycraft, who came to Birmingham in 1881 to become the school's first professor of physiology, but had by this time moved to Edinburgh.[29] The highest number of rejections was in midwifery (30 per cent) and chemistry (35 per cent).[30]

Course loads were also on the increase. The commencement of the five-year curriculum in 1892 witnessed the introduction of new courses in psychiatry, ophthalmic surgery, infectious diseases and the biological sciences. While this last requirement had originally inspired the merger with Mason's College, the former subjects encouraged schools to affiliate more closely with specialist hospitals in order to provide students with the appropriate clinical cases. For example, according to rules set by the College of Surgeons, teaching on the diseases of the eye required every student to hold a dressership in ophthalmology in a recognised specialist

[24] Ibid.

[25] UBSC, Mason's College, Medical Department, Dean's Register of Students, 1892–1929.

[26] M. Taylor, *Sir Bertram Windle: A Memoir* (London, 1932), 33.

[27] Ibid., 38.

[28] UBSC, Mason's College, Medical Department, Dean's Register of Students, 1892–1929.

[29] W. J. O'Connor, *Founders of British Physiology: A Biographical Dictionary, 1820–1885* (Manchester, 1988), 240.

[30] UBSC, Mason's College, Board of Medical Studies Minutes, 1892–1900.

department or at an eye hospital for a period of three months. Meanwhile, the subject's greater importance in the curriculum resulted in this particular lectureship being transformed into a professorship in ophthalmology by 1897. An isolation hospital where students could observe cases of infectious disease, on the other hand, had existed in Birmingham since 1875.

Other changes in the curriculum were less related to the dictates of the examining bodies in London. For example, in 1893, local ophthalmic surgeon David Charles Lloyd Owen wrote to Bertram Windle recommending the establishment of a chair in the history of medicine,[31] a post that existed at medical schools in Italy, Germany and Austria, but not in England.[32] Though the suggestion was supported by the editors of the *Lancet* earlier, it was not supported by medical school staff on this occasion, though Lloyd Owen was permitted to provide the odd lecture as he had done at Queen's College. In 1898, another initiative was proposed by Joseph Chamberlain, who was eager to commence instruction in tropical medicine for colonial medical officers as existed at Netley and Haslar to medical officers of the army, navy and Indian Medical Services. At the time, almost a fifth of medical graduates went on to practise in foreign climates, and many more would presumably do so given the unsettled political environment in South Africa. Despite these circumstances, there was, not surprisingly, some hesitancy to support such a programme in landlocked Birmingham, perhaps more suited to training 'captains of industry', as opposed to empire.[33] As such, a proposed school for tropical medicine did not appear in the Midlands, but two others were quickly established in London and Liverpool in 1899. Instead, the faculty regarded it as far more desirable to found a school of pharmacology, run according to the requirements of the Pharmaceutical Society of Great Britain. A pharmaceutical course had been considered as early as 1892 in order to improve the medical students' knowledge of crude drugs, their processes of manufacture, composition and doses. As importantly, this would provide students with additional access to pharmaceutical apparatus, as existed at Owens College in Manchester, Birmingham's greatest rival.[34]

Like Owens College, which became England's first civic university in 1880, Birmingham's educational venture eventually evolved into a university by 1900 and, incidentally, retained Mason's trademarked mermaid in its

---

[31] Ibid.

[32] F. Huisman and J. H. Warner, 'Medical Histories', in *Locating Medical History: The Stories and their Meanings*, ed. J. Huisman and J. H. Warner (Baltimore, 2004), 6–12.

[33] UBSC, Mason's College, Board of Medical Studies Minutes, 1892–1900.

[34] Ibid.

crest. Finally, and rather appropriately, given the gender associations of this particular symbol, although many women had been practising in various capacities in the city's voluntary hospitals, only from this date did medical education in Birmingham open to women. However, this only partly explains the increase in student numbers that came with a university charter. These and other issues related to this new era in medical education will be addressed in detail in Chapter 8.

A REGIONAL VISION:
THE EYE HOSPITAL

As mentioned previously in Chapter 4, the Eye and Ear infirmaries went their own ways in the last years of the nineteenth century, with the eye charity experiencing a period of noticeably rapid growth. In 1885, with fifty-five beds, the new Eye Hospital treated almost 1,000 inpatients and 15,000 outpatients. It had formally opened its wards to students in 1886 and began to organise a museum and library for teaching purposes. Though these pupils would have seen the full range of specimens they might encounter in private practice, they progressively saw far fewer patients from more distant parishes. By 1881, Wolverhampton, which had sent so many of its serious ophthalmic cases to Birmingham in previous decades, had established a similar specialist eye institution; though opening with only seven beds, the Wolverhampton Eye Infirmary treated 2,264 cases in its first year, the number of beds at the hospital climbing to fifty by 1897.[35] Seven years later it had approximately 6,000 patients on its register and governors, as at many Birmingham hospitals, added 'Midland' to the infirmary's name.[36]

Unlike the Wolverhampton Eye Infirmary, the Birmingham Eye Hospital was a teaching hospital and its success could no longer be measured by patient numbers alone. A donation of £1,000 from Richard Middlemore in 1888 ensured that progress in the field of ophthalmology would continue to be promoted locally in a series of annual postgraduate lectures founded by the hospital's former consultant.[37] The first lecturer was D. C. Lloyd Owen, a surgeon at the Eye Hospital and honorary surgeon to the Children's Hospital.[38] Lloyd Owen's 1889 lecture on 'The Essentials of Ophthalmic Therapeutics' discussed the usage of various ophthalmic drugs, including miotics, which constrict the pupil and permit the

---

35 [Anon.], *History of the Wolverhampton and Midland Counties Eye Infirmary, 1881–1931* (Wolverhampton, 1931), 9–11.

36 Ibid., 37.

37 BCLLS, Eye Hospital, Annual Report, 1888.

38 Ibid.

drainage of intraocular fluids, mydriatics, such as atropine, which dilate the pupil, and cocaine, the pain-killing properties of which were discovered by the Austrian surgeon Karl Koller in 1884. Seven years later, the local anaesthetic was also in regular use at the Birmingham hospital, had obviated the necessity for general anaesthesia, as well as its dangers, in common ocular operations, such as iridectomy.[39]

Technological innovations also remained essential to this specialism and were comprehensively incorporated into hospital routine during the last decade of the nineteenth century, beginning with a microtome that was donated to the charity's pathology department in 1895. The following year, this department, like the rest of the Eye Hospital, was electrified. Bacteriological investigation also commenced on site in 1897. X-ray apparatus was not purchased, but an arrangement was made with Dr John Hall-Edwards, the Moseley-born 'Surgeon Radiographer' to the General and Queen's hospitals, to take any required skiagraphs for 10s 6d at a time.[40] By 1899, the Eye Hospital was not just associated with the medical school in Birmingham, but it was added to the list of those institutions that were recognised for teaching purposes by the Royal College of Physicians and Royal College of Surgeons in London. Finally, Birmingham's links with the manufacture of optical goods should not be underestimated. Even before the foundation of its eye charity, Birmingham was a town whose inhabitants considered the medical aspects of eyes as a very serious business. For example, silver spectacles in use throughout the country from the eighteenth century bore the hallmarks of local makers, such as Joseph Willmore, Samuel Pemberton, J. Willmot, Francis Clark, John Parkes and Thomas Millington. In the 1830s, J. Gargory, an Optician of Bull Street, advertised himself as a supplier of 'the trade with the best gold-plated or solid gold, silver and fine blued steel eye-glasses and folding hand and temple spectacles equal to London manufacture' and issued gold and brass farthing opticians with trade tokens that the public could use like vouchers. In the first half of the twentieth century, numerous ophthalmoscopes were manufactured in the city, including that produced by Philip Harris & Co. Ltd, and some London optical manufacturers, such as J. & H. Taylor, had branches in Birmingham.[41]

---

[39] H. Eales, 'The treatment of cataract', *Birmingham Medical Review* 27 (1890), 258. Iridectomy is an operation that involves the part removal of the iris.

[40] For more on Hall-Edwards see E. H. Burrows, *Pioneers and Early Years: A History of British Radiology* (Alderney, 1986), 39–40, 112–15.

[41] I am grateful to Neil Handley, Curator of the British Optical Association's Museum, for this information on Birmingham's manufacturing role in optometry.

RUNNING TO STAND STILL:
THE EAR HOSPITAL

Unlike the Eye Hospital, the Birmingham Ear Hospital in 1883 found itself once again occupying temporary accommodation and dispensing to patients, this time from premises in Great Charles Street. Nationally, and in contrast to Birmingham's underperforming charity, the ear-nose-throat specialism was making significant professional gains in these years. Besides gaining the *Journal of Laryngology* in 1887, the first British society devoted to laryngology, the British Rhino-Laryngological Association was established the following year by Morell Mackenzie, founder of the world's first laryngological hospital in 1865 at 32 Golden Square, London.[42]

By 1890, the governors of the Birmingham charity, too, were finally able to build a hospital with inpatient facilities at Edmund Street. Oddly enough, it opened its doors to medical students even before staff admitted their first inpatient, students having been admitted since 1892.[43] From this period, every three months another student was appointed to undertake the duties of clinical assistant to outpatients. Two years later, staff and students, not to mention patients, finally obtained the benefits of ten beds for inpatients. This figure alone, however, does not give an indication of the work undertaken at the institution. By 1894, staff saw 5,929 registered patients, who made 18,027 attendances.[44] An increase in operations, which approached 400 in number, equally demonstrated that the resources of otologists were no longer limited to irrigation. Diagnosis, however, still depended largely on clinical acumen and experience. By 1897, beds had increased to thirty and staff treated more than 6,000 cases, a third of whom still came from outside Birmingham. With nearly half of these patients paying a registration fee of 3s 6d, finances were stable and the institution's board even commenced plans to enlarge the hospital;[45] electric lighting was installed in 1895. Like at the Eye Hospital, the board also commenced to organise a library. Unlike at the Eye Hospital, this collection was intended for the entertainment of patients, not the instruction of students.

[42] Gould, *History of the Royal National Throat, Nose and Ear Hospital*, 10.

[43] BCLLS, Ear and Throat Infirmary, Annual Report, 1892.

[44] BCLLS, Ear and Throat Infirmary, Annual Report, 1894.

[45] BCLLS, Ear and Throat Infirmary, Annual Report, 1897.

THE END OF CONTROVERSY:
THE WOMEN'S HOSPITAL

In the last decade of the nineteenth century, cases at the Women's Hospital, as in previous years, continued to accumulate more quickly than funds. In many ways, this was chiefly the result of its location. The institution's distance from the city centre alone ensured that, despite the hospital staff's many remarkable achievements, few medical students walked its wards. Unlike the smaller specialist hospitals in the city centre, it was less conspicuous to potential subscribers. While it had gained the support of the Sparkhill community, especially its friendly societies, it was less known among ordinary townspeople outside its immediate environment.[46] Chairing the hospital's annual meeting in 1894, the Mayor of Dudley claimed he had 'seldom seen a worse financial statement' and would bring the hospital to the attention of the inhabitants of the Black Country.[47]

Consequently, hospital staff devised more innovative ways of increasing hospital income in these years. For example, Lawson Tait suggested that registration fees as existed at other specialist institutions be extended to the inpatient department.[48] Moreover, he pushed for the development of dispensing work for women, given the way in which this controversial issue had advertised the hospital when first introduced. Others suggested the hospital could begin a provident scheme, as staff knew of many 'shabby-genteel people' who would be pleased to pay the small fee required.[49] As patients tended to be the middle-aged wives of working men, it was also suggested the institution receive a greater share of the Hospital Saturday Fund. More successful was their decision to appoint three paid canvassers who managed to double the hospital's subscription list in 1895. A bazaar held that same year raised another £430, while a Grand Charity Ball on 8 January 1897 added another £1,000 to the charity's coffers.[50]

The position of women at the hospital continued to consolidate in these years, and not only in the dispensary. For example, the resignation of Thomas Savage in November 1897 left Dr Annie Clarke the most senior member of staff at the hospital. While all medical officers continued to act as pathologists, by 1899, all pathological specimens requiring examination were to be sent to Mason University College. Though female staff at the Women's undertook less pathological work than at other hospitals, the

---

[46] BCLA, Women's Hospital, Annual Report, 1897, HC/WH/1/10/4.

[47] BCLA, Women's Hospital, Annual Report, 1894.

[48] Ibid.

[49] BCLA, Women's Hospital, Annual Report, 1897.

[50] Ibid.

administration of anaesthetics had a year earlier been placed in the hands of a qualified female practitioner, Dr Mary Sturge. Unlike in previous decades, such appointments had become routine and were no longer contentious. Nevertheless, the hospital still managed to conclude the century with a bit of controversy.

On this occasion, the controversy originated with comments made by Christopher Martin, one of the hospital's surgeons. Appointed to assist Tait in 1890, Martin sent a letter to the *British Gynaecological Journal* in the summer of 1898 suggesting that the inpatient department at Sparkhill was 'ill-adapted to its work'.[51] In particular, Martin claimed an old-fashioned dwelling in Birmingham was less suited to the needs of a women's hospital than was 'an old-fashioned farm house in the country'.[52] While its medical committee considered the hospital to be equal to if not better than any other women's hospitals undertaking abdominal sections ten or fifteen years previously, other hospitals had since made advances which are wanting or impossible in Sparkhill. In particular, Martin regarded the hospital's abdominal-section wards to be inconveniently small. Measuring 2,400 and 1,720 cubic feet, the wards held seven or eight persons during operations. As such rooms tended to be occupied by both patient and nurse for five to six days following operations, the atmosphere became 'very close and oppressive'.[53] Those patients lucky enough to be transferred from the operation ward to convalescent wards were no better off. Most were moved by two men on stretcher up a steep narrow winding staircase to the top wards of the building, the stretcher often being held at an acute angle on its ascent. According to Martin, this disturbance always alarmed and upset patients so shortly after their operations and often complications set in as a result of removal.

In general, most other defects stemmed from the hospital's layout. For example, the ventilation of the building was regarded as defective, with the top wards accumulating warm air from the lower floors. Opening windows, on the other hand, merely caused drafts. In the event of fire, the central staircase would allow a fire to quickly build, in which case the ten patients at the top of the building would be left quite helpless, the staircase being their only way out. Bedpans from these top floors were carried through the house and past the abdominal-section room; the atmosphere of the house was rendered offensive in this way and because the water-closets were not cross-ventilated. On a daily basis, major operations were carried out in five

[51] BCLA, Women's Hospital, Medical Committee Minute Book, 1893–1928, WH/1/5/2.

[52] Ibid.

[53] Ibid.

scattered rooms, two in the middle of the house and three in the 'cottage wards' outside. As a result, in most cases, surgical preparations were made around the patient, who saw all paraphernalia brought in and arranged, a situation that caused much nervous distress. All such preparations were undertaken by nurses, who carried the operating equipment from ward to ward. As a result, more modern, bulky operating tables could not be adopted, and staff had to be content with the 'old fashioned and primitive apparatus' that existed. As one might expect, after operations, the building's wooden floors became soiled with 'blood, pus and septic fluids'. These fluids became trapped between the planks and had the potential to cause septic problems later. Finally, the inpatient theatre was quite distant from the town. Consequently, in cases of severe haemorrhage much valuable time was lost, while other patients could not be seen by surgeons several times a day due to the institution's distance from town.

As one might expect, Martin's claims were contested by the hospital's first generation of medical officers as represented by Lawson Tait. To begin with, Tait insisted the hospital had greatly improved since he had ceased to be an active member of staff on 17 February 1893 and operations could now be performed out of the sight of other patients. While he agreed that patients should not be moved as was common at his own private hospital, he did not understand why men were required to carry women; at his hospital all such work was performed by nurses in carrying clothes in which no tilting was required. He was also hesitant to criticise the ventilation of the building, which he regarded as admirable. On the other hand, the dangers associated with fire he described as the most serious of the hospital's defects. These, however, he felt might be remedied by building a stone staircase from the front or rear of the building and so allow for the expansion of the wards where the old staircase used to be. In contrast, he remained opposed to the construction of an operation theatre as each patient already occupied their own rooms and preparations for operations could easily be made outside. The floors could also be made impermeable by painting, and only pine board and trestles, not expensive operating tables, were in his opinion required by the skilful surgeon. In general, he regarded developments in medicine such as cumbersome tables advantageous only to instrument makers. While the hospital's distance from the town was considered a negative aspect, Tait insisted the distance was more easily traversed than that between Harley Street and St Bartholomew's or the London Hospital and Cavendish Square. Neither did the journey increase the dangers associated with haemorrhage, he having been too late when living only two doors down from the hospital at no. 9 The Crescent. In general, Tait did not consider Martin's comments to be justified. Moreover, he considered the letter to be 'disloyal in an extreme degree'. Clearly, Tait was more irritated that he had heard no complaint from Martin, his assistant for nearly a decade, prior to the letter's publication. In reply, Martin suggested that even if the hospital's mortality were nil, it could still be ill-adapted. Siding with Tait, the general committee regarded Martin's article 'a grave censure on those responsible for the arrangements at Sparkhill', requesting that he add an explanatory note to the published text. Martin would later submit this note to the journal to be printed alongside his inflammatory letter.

As a result of the nature of the debates arising at this time, the death of Lawson Tait in 1899 truly marked the end of an era at the Women's Hospital. Evidently not entirely of Tait's opinion, staff finally devised plans for the extension of their specialist institution in the final year of the century. To begin with, the number of beds, distributed among ten wards (or rooms), was increased to twenty and operations, which numbered 384, were no longer performed among the other patients in the main building, but in a renovated cottage ward, equipped with aseptic lavatory arrangements and efficient lights. To the satisfaction of staff, especially Martin, the death rate following reconstruction was reduced to 2.94 per cent after averaging nearly 5 per cent for half a decade. That same year, the hospital rectified another notable omission. Previously reliant on private convalescent homes, the hospital established its own home for six patients at 'Woodfield', Showell Green Lane, Sparkhill on 17 April 1899. Run by Nurse Hobbs with the assistance of a single servant, the building remained inconspicuous, as objections from neighbouring tenants prevented the installation of a brass plate naming the institution.[54]

BUILDING ON TRADITION:
THE GENERAL HOSPITAL

Despite such important developments at other voluntary hospitals, the premier medical institution in Birmingham at the close of the nineteenth century, as at its outset, remained the General Hospital, and plans to ensure and enhance its reputation in the region gradually emerged. As such, in the early 1890s, the blueprints for a new 346-bed General had been drafted. The person appointed to lead the design stage of the project was William Henman, a Nightingale-inspired hospital architect, who belonged to a small, but growing group of late-Victorian, specialist hospital architects. Like Nightingale, Henman clung to the basic configuration of the pavilion plan, designed to offer maximum fresh air by cross-ventilation, ideas that had hardly changed since their introduction in the years following the Crimean War. When Henman won the competition to design the General Hospital in 1892, however, he initiated the radical move to

[54] BCLA, Women's Hospital, Annual Report, 1899, HC/WH/1/10/4.

change the pavilion design from a natural to a totally mechanical system of ventilation;[55] a similar decision to introduce an artificial means of air circulation at the town's new workhouse in 1852 seems to have inspired much of the local support for mechanical means of ventilation.[56] He also championed the new perception of hospitals as 'health manufactories'.[57] Less innovative was his opinion that hospitals in external appearance should demonstrate the importance of the work carried on within, many contemporaries regarding architectural embellishment as a waste of charitable money.[58] Most architects had to steer a difficult route between such overspending and designing a well-endowed building that inspired civic pride. The designs for the new General Hospital tended to the former, but it also attracted much praise. Most notably, Sir Henry Burdett, the former Secretary to the Queen's Hospital, who had, since resigning from his post in 1873, become the leading expert on hospital charity and founder of the Hospitals Association, regarded it as 'the most beautiful designed building ever dedicated to the purpose of a hospital'.[59]

Within a year of winning the competition to design the new hospital, Henman had dramatically upgraded his plans at the request of his Birmingham clients to provide a fully mechanically ventilated solution using the plenum system, which relied on tempered air propelled around the building via ducts. Working with William Key, who had devised the plenum system at the Victoria Infirmary in Glasgow, Henman created a 346-bed design that consisted of a cross-ventilated corridor spine extending across the site, with four three-storey pavilions of twenty-four-bed wards arranged off the central building. Unlike previous hospitals in Birmingham, the windows of the new General Hospital remained sealed and were for the admission of light only. Relying on eight fans, from 6 to 8 feet in diameter, and driven by electric motors, the system heated and cleansed 20 million feet of air every hour before propelling it through about 35,000 feet of steam tubing, allowing the hospital air to be changed ten times per hour without opening any windows.[60] At the same time, it expelled an equal amount of air through its valved and louvred roof turrets. According to the editors of the *Lancet*, this new method of ventilation was 'far in advance

[55] J. Taylor, *The Architect and the Pavilion Hospital: Dialogue and Design Creativity in England, 1850–1914* (London, 1997), 16.

[56] R. Hodgkinson, *The Origins of the National Health Service: The Medical Services of the New Poor Law, 1834–1871* (London, 1967), 530–1.

[57] Ibid., 38.

[58] Ibid., 3.

[59] *Hospital*, 15 September 1894.

[60] Taylor, *The Architect and the Pavilion Hospital*, 9.

of any scheme which has hitherto come under our observation'.[61] Others were less impressed, and suggested an efficient system of ventilation might have been had for far less money by simply opening the hospital's windows, as was common practice at tuberculosis sanatoria across the country. While there was some concern that patients might leave hospital with the idea that the windows in their homes should not be opened when housing the ill, the mechanical system of ventilation fell out of favour among hospital administrators because it increased coal consumption fourfold. Not surprisingly, within only a few years Henman's designs for other voluntary institutions had again adopted more traditional, less high-tech and less costly methods of ventilation.[62]

Although many of Henman's ideas concerning hospital design would appear outdated only a few years later, the institution was, rather predictably, hailed as a showpiece of provincial medicine when it opened. Above all, and unlike the old building, it was in all respects suited to a city of half a million inhabitants. So long restricted in its growth due to the existence of other voluntary hospitals, the massive project would produce a hospital worthy of one of the nation's key industrial centres. Unintentionally, the very scale of the project also reduced the funds available to many other local medical charities and institutions, including the new university.[63] Equally, its comprehensive nature further developed specialist services, including dental, eye and ear clinics, already offered at the city's other medical institutions. Some co-ordination between voluntary hospitals was considered, but medical services were not seriously consolidated in the same way higher education was reorganised in these years. In subsequent decades, far greater efforts were therefore made to co-ordinate fund-raising campaigns and the success of these alone ensured that the voluntary model of hospital care would survive largely unchanged throughout much of the next century.

Ultimately, the strength of local industry would permit the city's other voluntary hospitals to rebuild in the first decades of the twentieth century. As teaching hospitals, however, these medical institutions had little choice but to spend huge sums of money in order to remain cutting-edge and maintain their hard-fought national, if not in some cases international, reputations. As many quickly learned, elevated status also significantly inflated the costs of health care. Financial crises, as in the past, always loomed and would regularly punctuate the next half-century, but less so than in other provincial centres. Besides possessing a strong economy, Birmingham

[61] *Lancet*, 11 May 1895, 1203–5.

[62] Taylor, *The Architect and the Pavilion Hospital*, 196.

[63] E. Ives, D. Drummond and L. Schwarz, *The First Civic University: Birmingham 1880–1980: An Introductory History* (Birmingham, 2000), 75.

pioneered new charitable methods, which allowed its medical charities to tap into the region's relative prosperity in order to maintain these institutions and the important work they undertook. While it has been argued that hospital annual reports in nineteenth-century Birmingham constantly reprinted scarcely varying subscription lists comprising the names of some 2,000 philanthropic individuals and families, Hospital Saturday, in subsequent decades, evolved into a more comprehensive contributory scheme that effectively democratised charity and linked most of the city's workplaces to their health care services. Music festivals may have declined, but these gave way to charitable sporting events, among a host of other popular leisure activities similarly linked to medical fund-raising. Together, this played no small part in ensuring the survival and further promoting the growth of Birmingham's hospital services, as well as maintaining the status of a city that had, over the course of approximately a century, been transformed from the 'toy-shop of Europe' to England's Second City.[64]

[64] Hopkins, *Birmingham: The Making of the Second City*.

PART II

The Teaching Hospitals
in the Twentieth Century

1900–1939

CHAPTER 7

Reconstruction Delayed: The Development of the Teaching Hospitals, 1900–1919

A T THE THIRTY-NINTH annual meeting of the governors of the Children's Hospital held at Birmingham Council House in the first year of the twentieth century, the Lord Mayor, Alderman Edwards, attempted to deflect some attention from the financial crisis facing the institution's management. He claimed that when the history of the Victorian period came to be written 'there would be no more fascinating chapter than that which dealt with the emancipation of childhood and the recognition of the claims of women in the world'.[1] For the time being, however, the town's inhabitants heard little else than an 'annual wail' from hospital governors as their subscription lists continued to shrink.[2] More observant subscribers that year would have noticed that the institution's income was higher than ever due to an unusually large number of timely legacies. In less than a decade, however, numbers of subscribers declined by 15 per cent, down from 1,039 in 1894 to 878 in 1901. Similar declines are noticeable in the annual reports of the city's other voluntary hospitals, not to mention in those issued by the nation's other 600 voluntary hospitals. These, and many other changes affecting hospital finance, not surprisingly, made it difficult for governors at all of Birmingham's teaching hospitals to contemplate reconstruction in these years. An infant mortality rate that averaged 157 per 1,000 births in the first five years of the century, on the other hand, went some way towards convincing many hospital administrators that only greater efforts were required.[3]

BRICKS AND MORTAR:
THE CHILDREN'S HOSPITAL

Unlike the newly built General Hospital, Heslop's hospital for children appeared out-of-date in the first decade of the twentieth century. According to visitors, a 'feeling of depression came upon one in going over it'.[4] Others claimed hospital accommodation required immediate improvement or it would become more difficult to attract staff to fill vacant posts. Despite

---

[1] BCLA, Children's Hospital, Annual Report, 1900, HC/BCH/1/14/7.
[2] BCLA, Children's Hospital, Annual Report, 1901, HC/BCH/1/14/7.
[3] Hopkins, *Birmingham: The Making of the Second City*, 150.
[4] BCLA, Children's Hospital, Annual Report, 1907, HC/BCH/1/14/8.

133

such warnings, when Miss Bell resigned as matron in 1908, sixty candidates applied for the post, which went to Miss E. H. Grime, formerly of Dewsbury General in Yorkshire.[5] Though trade in Birmingham remained depressed in the early twentieth century, the hospital's governors decided to 'take the plunge' and commence reconstruction, as it was believed the public would not abandon them.[6] Though the hospital's lease had another ten years to run, the governors had their eyes on a site in Ladywood Road that was well connected by tramway to other parts of the city. As a result, the site was acquired and plans for redevelopment commenced, only to be delayed by a predictable shortage of funds.

By 1910, plans for the new hospital revived when the *Birmingham Daily Mail* opened the King Edward VII Memorial Fund.[7] Intended for a statue to commemorate the late king, the remainder of the fund was handed over to the hospital governors to support their institution. As a result, the charity was referred to as the King Edward VII Memorial Hospital thereafter, and construction of a 120-bed hospital commenced in the spring. By 1911, £35,000 had been spent on the new hospital, but another £15,000 was required by the time building commenced.

Medical practice at the hospital also changed noticeably in these years. For example, by 1902 operations increased by nearly 40 per cent to 2,191 in number from 1,593 four decades earlier; surgical costs rose by £96 accordingly.[8] The charity's beds, however, were not filled with surgical cases alone. Previously held for at least a day, patients undergoing operations of the nose and throat were discharged within hours, five beds being set aside for such cases. A decrease in inpatient numbers was due to the fact that patients operated for adenoids were sent home the same day. Though circumcision had previously been the most common operation at the hospital, by 1909 staff were removing more adenoids than foreskins (636:625), and the number of circumcisions only continued to shrink.[9] In any case, surgery was becoming a far more routine procedure at all hospitals in these years.

On average, the hospital's average length of stay declined to twenty days in these years of high infant mortality. Inpatients also tended to comprise far more serious cases, which explains a death rate of 15 per cent, at a time when Birmingham's other hospitals averaged less than half this figure. On the other hand, the outpatient department had become even more crowded,

5 BCLA, Children's Hospital, Annual Report, 1908.

6 BCLA, Children's Hospital, Annual Report, 1907.

7 BCLA, Children's Hospital, Annual Report, 1910.

8 BCLA, Children's Hospital, Annual Report, 1902, HC/BCH/1/14/7.

9 BCLA, Children's Hospital, Annual Report, 1909, HC/BCH/1/14/8.

especially on Monday afternoons, when clinics for school children were held. The medical inspection of school children threw a great deal of additional work on hospital staff. Begun in 1908, following the Education (Administrative Provisions) Act of 1907, the examination of children in day schools was introduced by Campbell-Bannerman's Liberal government, which commenced a string of important reforms in these years. Described as a major step towards the creation of the welfare state, which provided much new information about child health, to hospital staff the Act meant an increase in the number of outpatients.[10] More specifically, operations on adenoids increased markedly, reaching nearly 900 annually, the majority of local authorities concentrating on similar treatment, along with orthopaedic services and treatment for ringworm.[11]

Though hospital medical officers no longer made home visits, links with discharged patients were strengthened in an effort to extend the benefits of hospitalisation under a new programme of home visiting. Comprising five women and twelve visitors, the hospital's Aftercare Committee began work in February 1911, when seventeen children received weekly visits, as well as milk, eggs and 'good air'.[12] Visitors also sought to transform families' hygiene habits. For example, voluntary staff on one visit found a patient 'perched on the boundary wall of the family home so she might receive "better air"'.[13] The mother of another patient walked to a friend's home each day in order to obtain superior milk and Welsh water, ensuring her child would receive a diet similar to that prescribed in hospital. Another was found 'in a broken bath in the centre of a dismal court where the family lived with a ring of brothers and sisters looking after her there instead of in the terrible place they called home'.[14] In an effort to avoid duplicating services, all cases receiving assistance were registered at the City Aid Mutual Registration Office and reported in hospital annual reports.

By 1913, readers of the institution's annual reports also learned that the new Children's Hospital had advanced beyond the planning stages. On 23 April, Princess Louise, Duchess of Argyll, the childless fourth daughter of Queen Victoria, laid its foundation stone, though £9,000 was still required to complete the inpatient department free of debt. Outpatient facilities would cost another £10,000, but the hospital's two departments

[10] B. Harris, *The Health of the Schoolchild: A History of the School Medical Service in England and Wales* (Buckingham, 1995), 1–3.

[11] Harris, *Health of the Schoolchild*, 4.

[12] BCLA, Children's Hospital, After Care Sub-Committee Minute Book, 1911–34, HC/BCH/1/9/2.

[13] BCLA, Annual Report, 1911, HC/BCH/1/14/8.

[14] Ibid.

would again share the same site after nearly fifty years of separation, thereby reducing running costs. The change of premises alone reduced staffing costs, and outpatients no longer waited on the street in front of the hospital.

While the King Edward Fund raised approximately £30,000 for the hospital, novel methods were devised in order to make up the deficit in the building fund. The most innovative was conceived by Mrs J. E. Player, who organised a novel brick-laying ceremony in July, when 480 guinea bricks were sold to the public, eighty-two being purchased by various schools in the city, including fifty-nine Council schools.[15] At the ceremony, hundreds of children laid bricks marked with their individual initials, each junior bricklayer receiving an engraved souvenir trowel. At the hospital's opening, a register of bricks was placed in the entrance hall permitting children to locate their personalised donations. In 1914, a further eight branches of the Brick League were commenced, many more opening in subsequent years and fostering additional competition among the hospital's youngest donors.[16]

Instead of relying on random gifts, as in the past, another initiative involved printing those items the hospital needed most in the pages of annual reports. Additionally, the hospital provided material to children, mainly school girls, who made clothing and bedding in needlework classes. In its first year, the scheme provided material for forty cots (1,320 items).[17] At meetings in Edgbaston, children sewed bibs, vests and bed-jackets. In 1915, the hospital issued pocket sewing cases containing paper patterns for those items immediately required by the charity. A year later, materials were pre-cut by a needlework department in Council House and distributed to schools, thereby rendering the employment of two permanent sewing maids unnecessary.[18]

Despite much community support, the hospital's reconstruction eventually slowed due to a labour shortage brought on by war. Nevertheless, the hospital opened on 24 December 1916, though without official ceremony. Little fanfare accompanied a second royal visit in May 1919, when the king and queen viewed the building and its 150 beds in 'normal working condition'.[19] The hospital remained in an unfinished state, reconstruction lasting the duration of the war. Two additional wards of twenty-five beds

[15] BCLA, Annual Report, 1913.

[16] BCLA, Description of Function and History of Brick League, 1932, HC/BCH/1/13/11.

[17] BCLA, Annual Report, 1913, HC/BCH/1/14/8.

[18] BCLA, Annual Report, 1918, HC/BCH/1/14/9.

[19] BCLA, Annual Report, 1919.

each opened in 1918 and an outpatient department and two wards were still in the process of completion at war's end. Once finished, the new hospital provided three times the accommodation of the Broad Street building, though working costs only doubled.

THE RISKS OF INFECTION: THE ORTHOPAEDIC HOSPITAL

Regarded as Birmingham's second children's hospital, the Orthopaedic Hospital commenced the twentieth century with a deficit which averaged £500 for nearly a decade. Still the only hospital of its kind outside of London, the work of the charity only increased, though its premises were yearly becoming less suited to surgical cures. Inpatients numbered 912 in the first year of the century, while outpatients made more than 10,000 attendances, 7,000 patients attending for massage and electrical therapy.[20] While hospital surgeons operated upon nearly 500 patients annually, most deformities were still treated post-operatively with boots and splints, which remained the charity's single greatest expense (£683). Governors reduced spending in 1903 when the hospital began to train its own nurses, and proposed the appointment of an instrument-maker in 1905.[21]

Given the centrality of outpatient services to the charity, it was on the improvement of this department that governors concentrated at this time. Estimated to cost £4,000, construction commenced in 1905 and ensured a deficit for many years;[22] the absence of a member of the Freer family on the hospital's staff for the first time in nearly a century was another noticeable deficit (see Chapters 2 and 6). Despite this loss, the hospital had made some new allies, including Richard Cadbury, who bequeathed the hospital £2,250 that same year. Greater efforts were made to canvas the inhabitants of rural parishes, nearly half of patients having come from outside Birmingham. Two years later, the hospital's annual report announced the completion of a new (electrified) outpatient facility and a £2,800 mortgage.[23] Set only to increase, outpatient attendances surpassed 12,000 in 1908. In contrast, inpatient numbers plummeted since the first year of the century, averaging only 300 for several years. Though convalescent beds at Emscote – donated by the Countess of Warwick in 1901 – and Moseley Hall were intended to boost inpatient numbers, the regular, if unintended, admission of infectious cases greatly hampered the work of the charity and threatened to transform it into a dispensary once again.

[20] BCLLS, Orthopaedic Hospital, Annual Report, 1901.
[21] BCLA, Orthopaedic Hospital, General Committee Minutes, 1901–6, HC/RO.
[22] BCLA, Orthopaedic Hospital, General Committee Minutes, 1905.
[23] BCLA, Orthopaedic Hospital, General Committee Minutes, 1907.

An outbreak of diphtheria in 1909 alone led governors to close the institution for five months, and repeat outbreaks severely restricted visits to patients, who often remained in hospital for as many months. Clearly, an outpatient department was only one of the hospital's many requirements in these years.

The years prior to the outbreak of war brought little respite. Besides reporting the death of Charles Warden, the annual report in 1912 revealed a decline in much-needed subscriptions. The only gains related to patients, their numbers rising due to a polio epidemic sweeping the country that year. Consequently, 'nervous diseases', accounting for 18 per cent of cases, outnumbered all other afflictions in 1912; the gradual separation of such conditions from other internal diseases continued in subsequent years, given a growing interest in the effect of industrial life on the human frame as well as an understanding of the nervous system more generally. Expansion was again discussed by governors, who decided to publish photographs of their patients in annual reports in order to encourage greater acts of charity. Scheduled to coincide with the hospital's centenary, the proposed reconstruction attracted some support, but was postponed by the outbreak of war.

A SOUND UNDERTAKING: THE EAR AND THROAT HOSPITAL

Extensions were due at other specialist hospitals and planned with equal caution. At the Ear and Throat Hospital governors opened a building fund in 1900, but delayed a public appeal to augment their thirty-one beds. A year later, only a tenth of the projected £6,000 had been collected, as grounds for reconstruction were not as strong as at other hospitals. According to hospital administrators, because the nurses were all 'real ladies', new accommodation was required to house them according to their status.[24] Expansion also allowed the caretaker to live on site and created room for isolation wards. An outbreak of scarlet fever in 1902 affected 5,000 of the city's children and resulted in the closure of the hospital's main ward. Though the institution was still the only provincial ear hospital outside Shrewsbury, Hereford and Bristol, the hospital's governors made the unusual decision to restrict entrance to those living within 5 miles of the institution.[25] Though this policy hindered subsequent fund-raising initiatives, the extension fund reached £1,293 in 1902. As a result, the board undertook their planned development, employing the architects Messrs Cossins, Peacock and Bewlay to direct alterations.

[24] BCLLS, Ear Hospital, Annual Report, 1901.
[25] BCLLS, Ear Hospital, Annual Report, 1902.

During four months of construction the inpatient wards remained closed. They reopened in 1903 with forty-one beds and an isolation block for measles and scarlet fever. While the hospital treated many children, this too was changing, a ward for adults reducing the number of children's cots. Interestingly, the majority (85 per cent) of 781 inpatients at the hospital in 1903, suffered from afflictions of the nose, and only seventy-five, thirty-one and ten individuals reported with diseased ears, throats and various 'other' ailments respectively.[26] Despite transforming a consulting room into an outpatient operation theatre in 1904, these proportions remained unchanged for the next decade, when ear cases climbed slightly at the expense of throat afflictions. Nevertheless, diagnosis of the latter cases improved considerably in these years. With the discovery of the tubercle bacillus and x-rays in the Victorian period and the introduction of the Wassermann reaction for syphilis in 1906, staff more easily distinguished between simple chronic laryngitis, syphilis, tuberculosis of the larynx and malignant disease. As such, surgical treatment of the larynx was set to enter a 'heroic' period.[27]

The appointment of an additional anaesthetist encouraged a steady rise in operations at the institution. During 1903, 866 administrations of anaesthetics are recorded. Additionally, ethyl chloride was used 755 (87 per cent) times, while chloroform was administered on thirty (6.2 per cent) occasions.[28] Over the next decade, ethyl chloride use remained steady, but chloroform grew in popularity, being used in a fifth of operations. In comparison, nitrous oxide was the preferred aesthetic at the Dental Hospital in these years, comprising 2,932 (94 per cent) of 3,108 administrations.[29] At the General, the anaesthetics used by medical staff varied a bit more over 2,528 administrations in 1902, with chloroform used 700 (28 per cent) times, then ether (647, or 26 per cent), and nitrous oxide on 368 (15 per cent) occasions.[30] The same was the case at the Queen's Hospital, where local anaesthetics were used more frequently than chloroform and ether, though the more frequent use of anaesthetics at all medical institutions in general led to further increases in surgical expenses all round.

Before the Ear Hospital's extension, the institution's deficit totalled less than £100. The extension, however, cost £9,000, a third more than planned, and collections raised only £2,016. As a result, £1,922 of the charity's investments were sold in 1903 and administrators negotiated a £2,200 overdraft

---

[26] BCLLS, Ear Hospital, Annual Report, 1903.

[27] Stevenson and Guthrie, *History of Oto-laryngology*, 128.

[28] BCLLS, Ear Hospital, Annual Report, 1903.

[29] BDHCL, Dental Hospital, Annual Report, 1900.

[30] BCLLS, General Hospital, Annual Report, 1902.

to cover liabilities.[31] At a time when subscriptions were declining at other voluntary hospitals, only £602 of the institution's assets remained invested. Consequently, although the charity was now able to offer painless aural treatment to many more local patients, the hospital's administrators would continue to live with the agony of debt for some years.

FINANCIAL FORESIGHT: THE EYE HOSPITAL

Staff and administrators at the Eye Hospital were also in favour of reconstruction in the early twentieth century. Unlike those of the Ear Hospital, its governors concentrated on improving sanitary arrangements at the institution. Work took nine months, the wards being closed throughout construction. More suitable accommodation for nurses and servants was also discussed, staff cubicles being 'far too cramped and small to be healthy'.[32] Unlike alterations at the Ear Hospital, these changes were not made for several years, and nurses were often reported sick during this period. More pressing was the need for a passenger lift and an outpatient department, which were configured by William Henman, the architect who designed the new General Hospital. Like the Midlands' premier teaching hospital, by 1903 the Eye charity had a new outpatient department and passenger lift, which was used to transfer patients to and from its new operating theatre and a refurbished chapel in the basement. The nurses did not receive their promised accommodation until 1908, when the former premises of the Girls' Friendly Society in Barwick Street were acquired. A tennis court and croquet grounds in Sir Harry's Road were donated by Mr W. A. Cadbury for the use of nurses at this and the city's other specialist hospitals.

Far more revolutionary changes were under way in 1909. In this year, it was decided to open four wards to paying patients, who were charged 2 guineas weekly.[33] Removing the hospital kitchen into the basement of the new wing, the building committee made room for sixteen additional beds, and the operating theatre was also enlarged. In total, these alterations and further changes to the nurses' home cost £10,000. The new wing was completed by 1910. Along with a greatly improved kitchen, the hospital had 110 beds, fifteen fewer than at the Manchester Eye Hospital, which underwent reconstruction the same year.[34] Its four private wards admitted nineteen

---

[31] BCLLS, Ear Hospital, Annual Report, 1903.

[32] BCLLS, Eye Hospital, Annual Report, 1900.

[33] BCLLS, Eye Hospital, Annual Report, 1909.

[34] BCLLS, Eye Hospital, Annual Report, 1910; Stancliffe, *Manchester Royal Eye Hospital*, 73.

patients in their first three months. Besides an improved x-ray department manned by Hall-Edwards and Emrys-Jones, the staff and patients benefited from a new heating system and hot water supply. Once again, the fortunes of the Ear Hospital were very different from those of the Eye Hospital, where finances kept apace with increasing patient numbers. By 1912, existing funds and demand encouraged the board to transform its first-floor offices into a department for outpatients, whose numbers, as in Manchester, surpassed 30,000 annually.[35] Besides x-ray apparatus and a dark room, the new department, and the rest of the hospital, benefited from electric lighting.

INFANTS AND INFANTRY:
THE DENTAL HOSPITAL

As the representatives of the first university dental school in the country, the Dental Hospital's governors convincingly outlined their case for a new hospital in 1901.[36] Existing as it did in a private home adapted for the purposes of a hospital, the building required complete redesign, not minor alterations. Designed by Messrs Bateman and Bateman, the proposed project was to cost £10,000–11,000, fittings and fixtures requiring another £4,000–5,000. Unlike other specialist hospitals, the premises were to be more elevated. Like that of the 'artistic jeweller', with whom they trained in a former era, the dentist's work required much light, and the hospital's third floor was to have the finest filling room in England.[37] The larger premises at Church Street and Great Charles would also improve sanitary arrangements, while also preventing any problems associated that came with mixing the sexes. Not only were separate waiting rooms planned, but patients would not meet when entering or leaving. At the time of the announcement, only £940 of the projected costs had been received. Unlike at other institutions, the timing of these alterations was well planned. According to hospital governors, since the General Hospital's reconstruction in 1897, a number of other hospitals stood aside to aid the Dental Hospital's appeal. Should their canvas have been unsuccessful, warehouses had been incorporated into the building's design and let for a 'substantial sum' to make up any future deficit.[38]

By 1906, the new building was completed free of debt. It attracted many visitors and similar hospitals for Manchester, Liverpool and Sheffield were

---

[35] Stancliffe, *Manchester Royal Eye Hospital*, 73.

[36] UBSC, Dental Hospital, Annual Report, 1901.

[37] UBSC, Dental Hospital, Annual Report, 1902; L. Lindsay, *A Short History of Dentistry* (London, 1933), 63.

[38] UBSC, Dental Hospital, Annual Report, 1902.

soon discussed. Besides the installation of a telephone in 1907, annual costs remained stable at £2,000 until 1913 when the dental school was redecorated for £96. That same year, further alterations were made, including the installation of fountain spittoons in the conservancy room and the acquisition of a special chair for the administration of anaesthetics.[39] Radiological equipment, however, was yet to be acquired.

While the work of the hospital continued to increase, it was the amount of conservation work, rather than extractions, that increased. The state of Birmingham's mouths, however, had not improved, staff claiming that patients' teeth showed considerable deterioration compared with fifty or a hundred years earlier. This was attributed to changes in diet, though other considerations included the consumption of alcohol, unwholesome foods and deleterious water, as well as the absence of phosphates in bread. While the disappearance of teeth was also linked to the survival of slum dwellings, the teeth of rich and poor were regarded as equally deplorable.[40] In 1908, hospital staff appeared even more concerned about the state of dental hygiene in the city. Unless some active measures were taken, governors predicted 'a progressive deterioration of the race'.[41] By 1907, extractions were again on the increase with many more individuals receiving professional dental treatment for the first time in their lives.

Even before the introduction of a school medical service in 1907, many more local children received dental treatment at the hospital. A particular donation of £1,200 in 1903, for example, permitted staff to examine the teeth of school children each afternoon. During the project's first three years, 1,400 school children at Summer Lane School had been treated, while the school's teachers attended a lecture on 'the care of children's teeth' and distributed toothbrushes.[42] The following year (1907), the medical inspection of children, including their teeth, was introduced by Act of Parliament. At the hospital's annual meeting that year, Viscount Cobham, while particularly keen to deny any Socialist connections, emphasised the importance of funding such schemes with public money. Of 2,733 mouths examined in 1907, 2,629 children required dental treatment, girls having been described in greater need of treatment, but also 'better patients'.[43] Only 4 per cent were found to have normal, healthy dentition. By 1910, when public health dentistry was only commencing in London, staff at Birmingham's dental hospital treated more than 5,500 children at five

[39] UBSC, Dental Hospital, Annual Report, 1913.
[40] UBSC, Dental Hospital, Annual Report, 1907–8.
[41] UBSC, Dental Hospital, Annual Report, 1906–7.
[42] UBSC, Minutes of Elementary Schools Subcommittee, 1906–12.
[43] Ibid.

schools.[44] At Summer Lane, where 2,389 children were examined, staff performed 7,120 operations, and students made 2,185 attendances. The other schools (and their associated student, operation and attendance figures) were: Bristol Street (1,310; 5,563; 1,613), St George's (718; 2,129; 739), St Chad's (716; 953; 314), and the local Hebrew School (439; 633; 194).[45] At the end of the year, the funds to support this work were exhausted and the school scheme was suspended. All work in connection with school children was subsequently taken over by Birmingham's Education Department at rooms which they rented from the dental charity for £100 per annum. The educational authorities eventually vacated the hospital in 1917 following the completion of their own dental offices.

The work of the hospital, however, did not abate despite greater state involvement in health care. By 1914, staff commenced a considerable amount of war work, 7,364 operations being performed that summer on army recruits, who would otherwise have been unfit for military service.[46] Perhaps conscious that 3,000 men had been invalided home during the Boer War because of their teeth,[47] governors subsequently claimed the hospital had provided the army with an additional battalion, 1,000 men having been rendered fit as a result of dental treatment. So, too, was the dental treatment of consumptive patients arranged, the care of teeth having been regarded as central to the eradication of tuberculosis; rarely regarded as life-threatening, dentistry was more likely to be included in discussions of public health in the twentieth century as the focus of reformers began to shift from sanitation to personal hygiene. The provision of these and other services in these years led staff to describe the hospital as the 'centre of dental treatment in the Midlands'.[48]

SEEING THE LIGHT:
THE SKIN HOSPITAL

The development of the Skin Hospital in this period was not much different from that of the Dental Hospital. Commencing the new century with subscriptions totalling the 'ridiculous … sum' of £500, hospital staff limited themselves to updating the building's drainage work. Consisting entirely of businessmen, the hospital board had always run their charity prudently and avoided debt. Nevertheless, the development of new therapies and the

[44] Smith and Cottell, *History of the Royal Dental Hospital*, 71.

[45] UBSC, Dental Hospital, Annual Report, 1909–10.

[46] UBSC, Dental Hospital, Annual Report, 1914–15.

[47] J. Welshman, 'Dental health as a neglected issue in medical history: the School Dental Service in England and Wales, 1900–40', *Medical History* 42 (1998), 308.

[48] UBSC, Dental Hospital, Annual Report, 1918–19.

need to enlarge hospital premises made this record increasingly difficult to maintain.

In 1900, hospital staff treated only 181 inpatients, but outpatients numbered 6,000, of which 5,000 paid for treatment. Nearly 75 per cent of all outpatients were skin cases, and 57 per cent of these were women. The remaining 25 per cent of outpatients were urinary cases (1,559), comprising nearly 85 per cent men, many suffering from venereal afflictions.[49] Over the next two decades, however, these proportions changed almost as much as patients' treatments. Though more expensive, new treatments proved highly effective and reduced the average stay of inpatients, which had reached fifty-four days in 1902, allowing staff to treat greater numbers of 'respectable' men, women and children.

Treatment at the hospital largely comprised drug therapy. In addition, staff bathed approximately 200 patients annually and conducted a few dozen operations, including circumcisions, the excision of growths and spooning for lupus. Surgical treatment declined in 1901 with the introduction of the hospital's first x-ray apparatus and an improved Finsen lamp, which greatly enhanced the treatment of cases such as TB, rickets, smallpox, *lupus vulgaris* (tuberculosis of the skin) and wounds. Finsen therapy had been pioneered by Danish physician Dr Niels Finsen, who developed TB treatment using ultraviolet light in the closing years of the nineteenth century. A year before Finsen received a Nobel Prize for this work in 1903, only 103 Birmingham patients attended the hospital's electrical department. Radium treatment was also commenced at the institution before any other Midlands hospital. By 1905, the department treated 4,920 patients annually, and similar light departments were started at other skin hospitals nationally.[50]

Two years earlier, an increase in attendances of children suffering from eczema led staff to propose an extension of the hospital. The addition of a floor to the rear of the building took a year to complete and led the governors to draw on investments. Besides improving the building's heating and ventilation, governors renovated the operating theatre and doubled the dispensary's size. The premises and lighting were also electrified. A room set aside for pathological investigations in 1905 further aided treatment. For example, the discovery of syphilis's etiological agent, *Treponema pallidum* in 1905, the Wassermann test in 1906 and Salvarsan, the 'magic bullet' treatment (1909), enabled venereal disease to be identified and treated more effectively. New electrical equipment to treat ringworm was also introduced in 1905, allowing cases to be cured in three, as opposed to eighteen, months.

---

[49] BCLLS, Skin Hospital, Annual Report, 1900.

[50] Russell, *St John's Hospital*, 54.

By 1906, the hospital reported a deficit of £235, which would have been larger had not a nitrous oxide apparatus been donated to the institution. Other acquisitions were postponed, including additional radium, the miracle therapy, which staff suggested be paid for by imposing a tax on Midlands residents.[51] Instead, the hospital purchased a radium applicator from donations in 1909 and another was ordered in 1911. In order to cover the costs of new and expensive therapies, governors began to discuss a minimum income for hospitals at a time when the nation's politicians began to speak of a minimum wage for labourers.[52]

According to hospital governors, the ignorance of donors regarding the benefits of radium prevented their board from spending more on this 'great blessing' (see Chapter 9).[53] The use of tuberculin and vaccines, as well as Salvarsan, which was introduced in 1910, also increased expenses. The cessation of drug imports from the Continent during the First World War was another inconvenience staff would face in these years. Consequently, medicinal bathing, said to be 'out of fashion' in 1907, increased from 397 to nearly 1,000.[54] Still the only hospital of its kind within an 80-mile radius of Birmingham, the city's transformation into the nation's largest munitions area during the war ensured the hospital's workload would increase in subsequent years.

CARVING A NICHE:
THE WOMEN'S HOSPITAL

Though regarded as the premier centre for the medical treatment of Midlands women by 1900, the Women's Hospital had not been enlarged since its move to Sparkhill in 1878. A new site had been found in 1899, but negotiations broke off due to insufficient funds. As a result, the hospital continued to occupy two sites, which was costly, but also encouraged a certain degree of friendly rivalry. A single site remained contentious as not all governors preferred an institution in the country, some staff having favoured a hospital in town. Despite existing divisions, an appeal was launched by a united hospital board the following January.

To justify their venture, governors claimed the city's population had grown by 60 per cent since 1880, and many patients waited twelve to eighteen months for admission. Always well supported by women, the latest

---

[51] BCLLS, Skin Hospital, Annual Report, 1909.

[52] S. Pollard, *The Development of the British Economy, 1914–1950* (London, 1962, 32; S. Webb, 'The economic theory of a legal minimum wage', *Journal of Political Economy* 20, 10 (1912), 973–98.

[53] UBSC, Skin Hospital, Annual Report, 1913.

[54] UBSC, Skin Hospital, Annual Report, 1907.

appeal challenged men to pledge more than they had in the past; unfortunately, when hospital collectors visited local businesses, male managers were usually unavailable. When first founded, the hospital was opposed by many practitioners who felt women could be treated at the General Hospital. Unlike the General, the Women's Hospital was originally situated in a house adapted for the purpose of a hospital. Though successful cures were achieved, patients were moved with difficulty, and the yard wards were rarely used in poor weather. The nurses' accommodation was 'a disgrace to the Hospital'.[55] Nurses used the kitchen as a sitting room and dining room, and the lack of ward kitchens led to friction with servants. Formerly regarding nurses as mere machines, members of the hospital board advocated that nursing accommodation be moved away from patients and the 'smell of disinfectants'.[56] As rebuilding the existing premises was considered to be the equivalent of putting 'new wine into old bottles', staff desired a purpose-built hospital, even though the hospital's lease had another seventeen years to run.[57] As many landlords would not have them at any price, this was not straightforward. Despite strong prejudices, a new site away from the 'smoke of that large city' was acquired at Showell Green Lane. Recognising that hospital construction had advanced as much as the science of surgery, the governors wanted the best hospital in Britain, though no 'extravagant fads'.[58] Possessing £26,000, governors would begin reconstruction only when £30,000 had been pledged.

By 1903, construction of a fifty-bed hospital was in full swing. According to staff, the '[n]oise of adjacent building operations' made communication difficult.[59] However, the great pace of construction only hastened the stone-laying ceremony that year. The first stone was laid by Arthur Chamberlain, who reflected on the history of the institution he helped found. Reminding those gathered of a failed attempt by Dr Ingleby and Professor Berry to establish a special hospital for the diseases of women sixty-five years earlier, Chamberlain claimed Birmingham had the largest women's hospital in the kingdom.[60] Opened on 20 September 1905 by Mrs J. S. Nettlefolds, wife of the chair to the Management Committee and Chamberlain's daughter, eight of the hospital's fifty beds were for paying patients. If required, an administrative block could also accommodate twenty beds, and, given the building's design, a third floor containing twenty beds could easily be

55 BCLA, Women's Hospital, 1902, HC/WH/I/I0/4.

56 Ibid.

57 Ibid.

58 Ibid.

59 BCLA, Women's Hospital, 1903, HC/WH/I/I0/5.

60 Ibid.

added. More importantly, of England's seventeen women's hospitals, only two reported a lower death rate.[61] In contrast, it had accumulated the largest debts, totalling £8,000, and reconstruction continued. A pathology department, with its associated full-time officer, was still required and the hospital went without heating apparatus until 1908. Once completed, attention turned to the construction of a convalescent home, as the building adjoining the existing home at Woodfield obscured one side of the house. With room for twelve patients, the new home in Park Road levied a 10 shilling fee on patients residing beyond 15 miles of the hospital.

As planned, an outside visitor was invited to inspect the new buildings, the first to give his opinion of the hospital being Sir Alexander Simpson (1835–1916), the nephew of Lawson Tait's mentor, Sir James Young Simpson, and recently retired Professor of Midwifery at the University of Edinburgh. Local residents were also encouraged to visit the premises, in an attempt to generate publicity and additional charitable contributions. Unlike other hospitals, where access was strictly limited, often to a couple of hours each week, governors welcomed anyone wishing to view the building, as construction had generated a £1,800 deficit.[62] The only guests not in attendance were students, whose exhaustive exam schedule limited their appearance at the institution. Its location more than 2 miles outside the city centre made it very difficult for any students to attend the institution, as Simpson noted following his 1908 visit.[63] His report also reveals the hospital treated more than just gynaecological complaints as originally stated in the charity's remit; renal and pelvic cases had also been seen in great numbers. Neither were those with afflictions of the breast being sent to the General Hospital as formerly. The report concludes by encouraging instruction at the institution, especially postgraduate training. In marked contrast to hospital governors, Simpson suggested gynaecological cases could be used in teaching 'without distress to the patients', women overlooked on clinical rounds often asking the nurses why their cases were of no interest to medical staff.[64] According to Simpson, '[t]o be made the subject of a clinical demonstration is a sure guarantee to a hospital patient that her case will be thoroughly gone into and carefully managed throughout its course'. A strong culture of clinical teaching was also encouraged to keep staff abreast of medical advances.

---

[61] BCLA, Women's Hospital, 1904, HC/WH/1/10/5.

[62] For more on institutional visiting see G. Mooney and J. Reinarz (eds.), *Permeable Walls: Historical Perspectives on Hospital and Asylum Visiting* (Amsterdam, 2009).

[63] BCLA, Women's Hospital, 1907, HC/WH/1/10/5.

[64] BCLA, Women's Hospital, 1908, HC/WH/1/10/5.

Still the hospital's strongest supporters, local women started the Women's Hospital League in September 1907, branches forming in Handsworth, Sutton, King's Norton, Knowle, Packwood, Moseley, Northfield, Perry Barr, Solihull, Erdington and Barnt Green. The following year, the network raised £535 in new subscriptions and donations. With the help of four more paying wards in 1910, the hospital was finally able to invest some of its legacies. Reaching £1,869 in 1890, income had increased nearly fourfold in twenty years. Of £7,254 collected in 1910, funds came from a greater variety of sources: £1,289 (18 per cent) from subscriptions, £125 (2 per cent) from donations, £124 from charity boxes, £500 (7 per cent) from Hospital Saturday, £681 from legacies, £1,334 (18 per cent) from registration fees, £5 from grateful patients, £55 from the sale of instruments and bottles, £968 (13 per cent) from dividends, £17 from nurses' premiums, £16 from the dispensing pupils' premiums, £513 (7 per cent) from paying patients, £365 fees from patients living 10 miles distant, £141 sundry, £217 special donations, £137 from Hospital Sunday, £415 (6 per cent) from local friendly societies and £352 (5 per cent) from the Hospital League. Though the charity's balance sheet provides another good example of the voluntary sector's ability to augment and replace diminishing traditional sources of income,[65] its governors undertook improvements with considerable confidence in their prospects.

Besides investing more, governors now regularly planned hospital expansions. In 1909, a septic ward, named the Margaret Ward, added six beds to the hospital, which accommodated fifty-six patients. Additionally, following the death of Professor Taylor that same year, a home for women with cancer was opened in his memory, though the hospice was not officially connected to the hospital. A larger project was commenced in 1910, when the hospital governors decided to take over the Maternity Hospital and lying-in charity in order to ensure its survival. Had the hospital staff been 'eminently successful bonesetters or any other unorthodox healers' they would not have been subject to as much jealous dislike.[66] By 1911, both hospitals were governed by a general committee comprising twelve ladies and twelve gentlemen. The older institution took 'the younger under its wing and carry it along with the rest of its work'.[67] Taking only a few normal cases for the education of pupils, the charity treated primarily abnormal cases, including puerperal fever, which had been made notifiable. When two beds were found inadequate for these cases, five more were

[65] S. Cherry, 'Before the National Health Service: financing the voluntary hospitals, 1900–1939', *Economic History Review* 50, 2 (1997), 308.

[66] BCLLS, Women's Hospital, Annual Report, 1910.

[67] BCLLS, Women's Hospital, Annual Report, 1911.

added and septic cases from both hospitals were dealt with at Sparkhill. In April 1911, the hospital began taking all those sent by the local Health Committee and enlarging the present puerperal ward by adding eight beds, and another ten outside the hospital.

By 1912, the hospital had accumulated a £1,050 deficit, its maternity branch contributing another £1,600. With legacies declining from £4,000 to £90 in a year, a uniform system of accounts used by the Central Hospital Board in London was introduced to monitor spending. Additionally, the engineers Messrs Henry Lea and Son were consulted to advise on saving coal, water, gas and electricity. Determined to stamp out puerperal fever among the city's poor, the governors added fifteen beds to the Margaret Ward, funded largely by the local Health Committee. By spending more at Loveday Street on prevention, governors intended to save on cures at Sparkhill. An appeal for £6,000 was started to increase the Margaret Ward to twenty-five beds and erect an operating room. When enlarged, the ward housed 190 septic cases, 170 or more being puerperal sepsis, many sleeping on its veranda the whole year round. The year ended with the presentation of a convalescent home to the hospital (in memory of Mrs Leopold Myers, who had regularly attended the hospital's meetings). Furnished with twelve beds, the Myers home, located on 2½ acres of land at Cleeve Prior, in its first year took 256 patients, most coming from the Women's Hospital's septic ward and staying at least sixteen days.[68]

Despite the death of Arthur Chamberlain in 1913, hospital staff in these years did not concentrate on their losses. The year had also witnessed a 50 per cent increase in the work of the Maternity Hospital since joining the Women's. Governors claimed the charity allowed Birmingham's poorest women to be treated by the same physicians and surgeons 'as the richest women in the land'.[69] An improvement in the standard of the hospital's midwives was their next goal. For the time being, governors could only look forward to a day when people would say, 'She comes from Birmingham, she is alright.' To achieve such ends, the public's support, as always, remained crucial. In order to create 'Birmingham ... midwives they [would need] Birmingham money'.[70]

[68] BCLLS, Women's Hospital, Annual Report, 1912.
[69] BCLA, Women's Hospital, Annual Report, 1913, WH/1/10/6.
[70] Ibid.

AUGMENTING STAFF AND INFRASTRUCTURE:
THE QUEEN'S HOSPITAL

The first years of the twentieth century often witnessed the mechanisa-
tion of Birmingham's hospitals. At the Queen's Hospital this commenced
with the installation of electric light in 1899. Two years later, the installation of internal telephone lines improved communication between hospital
departments and, in 1912, extended to the police and fire services. The reach
of electricity was similar extended. The hospital's laundry was also mech-
anised in 1901, and, in 1912, a vacuum was purchased to clean the wards,
though the building continued to be heated by open fires. While technological change is said to have made work more hazardous, the regulation
of a large amount of machinery also created work, leading to the appointment of a firm of consulting engineers in 1906 and an engineer's assistant
in 1909.[71]

The first decade of the twentieth century also saw important additions
to the hospital's medical staff. Besides a third house physician, appointed
in 1900, two registrars were appointed in 1903 to ensure adequate record-
taking. A year earlier the work of the pathologist was transformed into
a separate post, and no longer delegated to house surgeons (see Chapter
6). Run by Stanley Barnes, the pathology department was simultaneously
refurbished at a cost of £500. In 1906, when the post was again vacant, the
pathologist's pay was doubled to £100 in order to attract an experienced
expert, though this did not have the desired effect.[72] By 1908, the holder of
the post was additionally recognised as visiting pathologist to the Women's
Hospital.[73] An 'expert' anaesthetist was similarly considered in 1912, house
surgeons having anaesthetised the majority of operative cases as part of
their work.[74]

Two years earlier, staff had also appointed the first non-white member
of staff. Having advertising the post of house surgeon in 1910, the general committee managed to attract only a single application, that of 'a
Hindoo gentlemen', G. V. Deshmukh.[75] Though this led some members
of the committee to suggest prolonging the deadline for applications, it
was decided to elect Mr Deshmukh following an interview. After further
discussion, '[t]his motion was subsequently withdrawn', and it was decided
to appoint Deshmukh for a period of six months when his appointment

[71] BCLLS, Queen's Hospital, Annual Reports, 1901, 1906, 1909.

[72] BCLA, Queen's Hospital, General Committee Minutes, 1896–1907, HC/QU/1/1/6.

[73] Ibid.

[74] BCLA, Queen's Hospital, General Committee Minutes, 1907–17, HC/QU/1/1/7.

[75] Ibid.

was reconsidered. Having performed satisfactorily, the surgeon was reappointed in September for another six months. Though Deshmukh does not make another appearance in hospital records, the contribution of Asian practitioners has since been recognised as invaluable to the provision of health care services in Birmingham and the rest of the United Kingdom, especially since the 1950s.

The hospital buildings were similarly augmented in these years. In 1904, governors purchased 107 to 117 Bath Row for £3,875 in order to add a nurses' home and a sixty-bed block for inpatients to the 132-bed institution. So as not to interfere with the reconstruction of the Women's Hospital, the £30,000 extension was delayed for two years, nearly two decades since the hospital's last appeal. Thereafter, work progressed rapidly, and a nurses' home opened in 1907. While all seventy nurses occupied rooms outside the hospital's main building, by 1909 patients literally resided out of doors, the top floor of the new inpatient block having been designed as an open-air ward by Arthur Foxwell, one of the hospital's physicians.[76] The first systematic attempt to use open-air treatment for non-tubercular diseases, the ward was judged a success in its first year, and staff commended nurses and patients who 'braved freezing wind and driving rain';[77] open-air treatment was subsequently introduced to similar praise at the new Children's Hospital five years later and on a greater scale at the General in 1918.

Though projected to have 192 beds, the enlarged institution had only 170, sixty of which were medical, 110 surgical. Besides a dispensary and a sixty-seat chapel, the new hospital block created enough additional space to separate the women's and children's medical wards. The hospital's original building, on the other hand, was used for surgical cases only. As a result, the hospital treated twice as many surgical inpatients, though medical outpatients surpassed surgical cases by 30 per cent.[78] More surgical cases necessitated a second operating theatre, as well as new sterilisation equipment and the appointment of a masseuse in 1908 to treat fractures. Among surgical inpatients, the main ailments treated were hernia (193), appendicitis (108) and carcinoma (104).[79] Interestingly, the surgical department's single greatest expense was rubber gloves, which in addition to a clean operative site and a mask were the only measures required to exclude infection; cleaning costs, however, are far more difficult to calculate. Gamgee tissue, developed at the hospital a generation earlier, on the other hand, was now hardly ever used by surgical staff. More often utilised was the

[76] *Oxford Dictionary of National Biography*, accessed 10 June 2006.

[77] BCLLS, Queen's Hospital, Annual Report, 1908.

[78] BCLLS, Queen's Hospital, Annual Report, 1910.

[79] BCLLS, Queen's Hospital, Annual Report, 1906.

radiography department, which was enlarged. The newly appointed radiographer, Mr F. Emrys-Jones, made 280 radiographs in 1909, 509 in 1910;[80] of those cases radiographed since his appointment in 1908, the majority were for fractures (84), embedded needles (48), suspected calculi (26) and dislocations (22).[81] The installation of new apparatus for the treatment of disease by radioactivity in 1910 only increased his workload and hastened a decision in 1912 that the radiographer attend the hospital three days a week. Not surprisingly, governors found the work of the department to be 'hurried' and appointed an assistant that same year. In 1913, the post was made full-time.[82]

Though admitting approximately 3,000 inpatients a year, outpatients approached 40,000, nearly 20,000 of which were accidents. The main casualties in 1908 were broken bones (524), burns (443) and bites (203).[83] The increase in all accidents in this and subsequent years was attributed to the electric tramway that was laid in Bath Row in 1906.[84] Transport issues also increased the work of those charged with cleaning the wards. According to the governors, '[t]he dry dust from pitch macadam (largely horse droppings) when the wind is blowing gets into the Buildings and is deleterious to the patients often necessitating the closing of the windows when they should be open'.[85] This was remedied in 1911 when both sides of the tramway on Bath Row were paved with wood. This left only complaints of increased noise, which was ameliorated in 1913, when the speed of trams passing the building was reduced to 5 miles an hour.

By 1911, the work of all voluntary hospitals was influenced with the passage of the National Health Insurance Act, which came into effect in 1913. Designed to protect workers earning under £160 from slipping into poverty through illness, the Act marks the first time the government directly intervened with positive intent in the personal health provision of the most vulnerable groups in society.[86] Like Hospital Saturday, the system operated on the lines of a contributory scheme, with workers contributing 4d a week, which was topped up to 10d by employer and government contributions.[87] The sums collected funded GP services for insured persons, as

[80] BCLLS, Queen's Hospital, Annual Report, 1909.

[81] BCLA, Queen's Hospital, House Committee Minutes, 1908–12, HC/QU/1/2/14.

[82] BCLA, General Committee Minute Book, 1907–17, HC/QU/1/1/7.

[83] BCLA, General Committee Minute Book, 1908.

[84] BCLLS, Queen's Hospital, Annual Report, 1909.

[85] BCLLS, Queen's Hospital, General Committee Minute Book, 1907–17, HC/QU/1/1/7.

[86] A. Hardy, *Health and Medicine in Britain since 1860* (Basingstoke, 2001), 45.

[87] H. Jones, *Health and Society in Twentieth-Century Britain* (London, 1994), 26–7.

well as tuberculosis sanatoria, maternity benefits and a Medical Research Committee (renamed the Medical Research Council in 1920). To prepare for its implementation, thirty-two Midlands hospitals, including the Queen's Hospital, held co-operative meetings to discuss the legislation. In general, it was believed the bill would lead to a reduction in donations, as well as beds. As predicted, governors reported the loss of fifty-eight subscribers, or £100 in subscriptions, and a decline in registration fees in 1913.[88] However, pressure on beds increased, with staff treating 7,218 insured people, most having been sent by their panel doctors. Consumptive cases also increased when the hospital was recognised as an institution for the treatment of TB under the Act. Besides remuneration from the local government board for this work, the hospital also attracted £200 in new subscriptions. Additionally, the Act encouraged governors to reappoint an enquiry officer to investigate the details of all insured patients and further improve record-keeping practices. While indicative of a shift in the relationship between voluntarism and the state, these changes simultaneously allude to the development of a local referral system.

THE HEALTHIEST OF CHARITIES: THE GENERAL HOSPITAL

As the sole hospital fully rebuilt in the last years of the nineteenth century, the General Hospital entered the twentieth century with few of the concerns shared by staff and governors at Birmingham's other medical institutions. Not only was it regarded as up to date, but there was room for expansion. Memories of the previous institution began to fade very quickly after 1901 when the old building was sold to the corporation and demolished.

As one would expect, given its recent construction, changes to the hospital were limited. However, by 1905, governors were authorising further work. For example, in 1901 balconies were attached to two children's wards. In the same year a new hot-water system with a central heater replaced one costing £1,263 that conducted steam to seventeen different heaters in the building and reduced coal consumption considerably. No longer regarded as satisfactory, the operating theatres were reorganised, the larger being divided into two small theatres with their own sterilising and anaesthetic rooms. Provided with special tanks for sterilised water, large sinks replaced the theatre's hand basins and students' seats were removed. A small theatre on the second floor was similarly altered, total changes costing £1,000.[89] While minor construction was more easily contemplated after 1905, when

[88] BCLLS, Queen's Hospital, Annual Report, 1913.

[89] BCLA, General Hospital, Annual Report, 1907, GHB 435.

the hospital received a £5,000 bequest, larger projects were commenced with more confidence. In 1908, the operating theatre used for the specialist departments was similarly remodelled and the nurses' home was extended to make forty-four extra bedrooms. Despite another timely legacy, the hospital's deficit at the end of 1907 rose from £7,698 to £10,061.[90] Concerned with its overdraft, the board commenced efforts to increase subscriptions, which numbered 2,000 in 1910. Neither were music festivals the answer to their financial difficulties, as was proved in 1909, when profits totalled £2,500, compared with £6,000 in 1901. Another event in 1912 raised £1,550, but cost £10,832, and subsequent festivals were cancelled due to the war. Paying beds were not considered, because of inadequate staff numbers.[91] Instead, governors encouraged donors to endow beds.

In these years, the General treated almost one-half of the patients attending Birmingham's voluntary hospitals. Besides consuming considerable funds, this work required a large staff, comprising largely nurses. In 1909, the appointment of fourteen additional nurses brought their number to 120. The following year, the hospital employed 127. With these increases, hours were finally reduced, leaving nurses with more time for eating and relaxation. Greater demand for massage and electrical therapy added to their duties, though much of this work was undertaken by students preparing to take the exam of the Incorporated Society of Trained Masseuses. Pressure on the institution had also resulted in some nurses bandaging casualty cases and stitching wounds, practices the hospital governors discouraged.[92]

With medicine growing more specialised, cases at the institution continued to overlap with those of the city's other hospitals. In 1901, staff treated nearly 2,000 dental cases and a second dental chair was considered.[93] The following year an aural surgeon and laryngologist, Dr F. W. Foxcroft, was appointed when the number of ear and throat cases began to increase. As at the London Hospital, Foxcroft was permitted to operate only through the nose and throat and through the ear unless cerebral or lateral sinus complications were present.[94] According to the annual report in 1902, staff treated more women and children than other Midlands hospitals, and not only as outpatients. In the old hospital, only seventeen beds and eight cots were reserved for special cases. By 1902, fifty-eight beds and thirty-one cots, and another twenty-one beds in side wards were reserved for special cases.

[90] Ibid.

[91] BCLA, General Hospital, Medical Committee Minutes, 1906–11, GHB 75.

[92] BCLA, General Hospital, Medical Committee Minutes, 1901–6, GHB 74.

[93] BCLA, Annual Report, 1901, GHB 435.

[94] BCLA, General Hospital, Medical Committee Minutes, 1901–6, GHB 74.

By 1906, another two adult beds and a cot were added with the establishment of a skin department, run by Dr A. Douglas Heath in its first three years.[95]

Though staffed with their own officers, the specialist departments greatly increased the work of hospital staff. In 1902, for example, the Resident Surgical Officer not only assisted the aural surgeon at operations twice a week and administered anaesthetics for the dental surgeon, but he was the sole officer in charge of gynaecological, ophthalmic and aural cases (nineteen beds).[96] As a result, an additional house surgeon was added to assist all specialist cases in 1903. In 1906, two more were appointed to assist the five existing house surgeons. The radiographer, John Hall-Edwards, also felt the strain of the specialist departments. Not only were x-rays used to treat lupus, rodent cancer and other forms of skin cancer, but a Finsen light had been added to his department in 1902. Each application of Finsen's apparatus additionally required the services of a nurse for two or three hours. Given its 'congested state', the department ceased taking dental x-rays in 1907.[97] In 1908, the total number of radiology cases jumped to 3,000, following a decision to treat ringworm, and the radiographer's hours were extended. In 1910, his work expanded yet again with a decision to hire radium two days a week for the treatment of cancer, gout, arthritis and dermatitis, though some cases had been treated previously using Hall-Edwards's personal supply.[98] In 1912, after devoting half of the department's budget (£200) to radium hire, governors decided to purchase their own supply, as well as two applicators. The following year the department produced more than 3,800 radiographs and treated 285 patients with 2,800 applications of x-ray and radium.

With an increase in the work of all specialist clinics from 1907, governors requested staff to discourage trivial cases from attending the outpatient department. Though 2,294 non-essential cases were attended the next year, all were conveyed to other institutions after receiving first aid. As many patients were transferred in 1909. Trivial cases subsequently declined at both general hospitals following the introduction of National Health Insurance, many patients being directed to their newly assigned panel doctors.[99] Additionally, from 1912, district nurses were involved in a scheme to attend patients, who required frequent dressings in order to save them repeat journeys to hospital. Nevertheless, thousands of patients continued

[95] BCLA, General Hospital, Medical Committee Minutes, 1906–11, GHB 75.

[96] BCLA, General Hospital, Medical Committee Minutes, 1901–6, GHB 74.

[97] BCLA, General Hospital, Medical Committee Minutes, 1906–11, GHB 75.

[98] Ibid.

[99] BCLLS, Queen's Hospital, Annual Report, 1913.

to frequent the hospital's new emergency ward, which opened with ten beds in 1913. A new pathological block was finished the same year, but was not be equipped until after the First World War.

RESOURCES AND RESOURCEFULNESS: THE ADDITIONAL CHALLENGES OF WAR

As one would expect, the outbreak of war in 1914 disrupted much work at the voluntary hospitals. While staff at the General spent an extra £1,000 on bandages, drugs and bedding at the outbreak of hostilities, only the most well endowed hospitals could stock scarce items at already inflated prices. From patients at the Children's Hospital who were unable to obtain the food ordered by its after-care committee to staff at the Dental Hospital who employed anaesthetics in only a tenth of cases (500), most hospitals reported shortages in these years.[100] Not all were as resourceful as the Women's Hospital, where staff, unable to obtain many of the drugs they normally dispensed, joined a Warwickshire initiative to grow medical herbs from seeds supplied by the Board of Agriculture.[101]

Funds were equally scarce, and demand for beds soon increased. At the General Hospital expenses in 1916 exceed income by £6,654 annually, as opposed to £4,428 the five years before war, and the deficit over three years was £20,007.[102] Some relief came from the war office, which paid £2,026 towards military cases, and the John Feeney Trust, which contributed £350 to off-set war-related inflation. Total patient numbers, however, now approached 50,000, 5,562 occupying beds. A request from the Local Government Board to open a department of venereal diseases in 1917 following the introduction of the Venereal Diseases Act only increased attendances by 500 the following year. Besides opening a similar department in 1918, the Queen's, which treated half as many patients as the General, opened an electrical and mechanical therapeutics department to treat nervous disorders, as well as joints and limbs to restore civilians and soldiers. The same year, 160 soldiers were admitted to the hospital, and military beds numbered 125 for a few weeks the next year.[103]

Civil wards remained crowded due to working conditions in local factories, especially munitions works. A subsequent increase in contributions from munitions firms brought about the first noticeable increase in subscriptions at the General in several years, but not all hospitals benefited

100 BCLA, Children's Hospital, Annual Report, 1917, HC/BCH/1/14/9; BDHCL, Dental Hospital, Annual Report, 1915.
101 BCLA, Women's Hospital, Annual Report, 1915, WH/1/10/6.
102 BCLA, General Hospital, Annual Report, 1916, GHB 437.
103 BCLLS, Queen's Hospital, Annual Report, 1918.

from such funding.[104] Drawing attention to the effects of munitions work, which was bound 'to tell upon the weaklings', Christopher Martin, surgeon to the Women's Hospital, urged local manufacturers to support the women's and lying-in charities in 1915.[105] Given that 'there was no city in the world where women were doing more work, and doing men's work, than in Birmingham', Martin suggested many of the hospitals' 5,000 cases could also afford to pay for treatment. Apparently, few did, and in 1917 an appeal was launched to clear the £15,000 debt accumulated at the Women's and Maternity hospitals. Already holding joint meetings and running a special venereal department from 1918, the general committee finally suggested incorporating the two charities.

Labour shortages were equally noticeable. Labour difficulties delayed the reconstruction of the Children's Hospital for much of 1915, but building resumed when local plumbers resolved their differences with the Heating Engineers' Union. While building work resumed, outpatients were restricted to forty each afternoon, as opposed to sixty-five previously, due to staff shortages, though no serious case was refused admission. At the Women's Hospital, outpatient work was reduced as much as possible in 1915.[106] Having disappeared from the medical staff at the Children's in the first decade of the twentieth century, female medical officers were again appointed when five male medical officers enlisted in 1914. As more young men joined the forces, junior residents also became scarce, though qualified women were not numerous enough to fill every vacancy. By 1917, 35 per cent of staff at the General and 17 per cent at the Queen's and Eye hospitals were away on service.[107] Nationally, the war led to the call-up of more than half the country's qualified medical personnel.[108] As a result, students occasionally took on additional responsibilities. At the Children's Hospital, G. E. Mullins, a fifth-year medical student, regularly performed the work of resident medical and surgical officers.[109] Other positions were filled, but only a month at a time, occasionally by foreign doctors.[110] By far the main difficulty facing the larger hospitals was a shortage of probationer nurses, their numbers declining by twenty at the General Hospital in 1918. At the

[104] BCLA, General Hospital, Annual Report, 1916, GHB 437.

[105] BCLA, Women's Hospital, Annual Report, 1915, WH/1/10/6.

[106] Ibid.

[107] BCLA, General Hospital, Medical Committee Minutes, 1916–21, GHB 77.

[108] B. Harris, *The Origins of the British Welfare State: Social Welfare in England and Wales, 1800–1945* (Basingstoke, 2004), 219.

[109] BCLA, Children's Hospital, Annual Report, 1916, HC/BCH/1/14/9.

[110] BCLA, Children's Hospital, Annual Report, 1917; General Hospital, Medical Committee Minutes, 1916–21, GHB 77.

Queen's Hospital, thirty-three 'special probationers' underwent a reduced period of training to remedy shortages in 1915. By discontinuing their private nursing scheme, the hospital did not face the shortages reported elsewhere.[111]

Though numerous, many war sacrifices went unreported, but were equally obvious. For example, by 1917, the Children's Hospital reduced their annual report given a shortage of paper, while only the governors at the General were supplied with printed reports in 1919. At the Queen's, all medical statistics, donations and legacies, as well as illustrations were omitted from reports in 1917, saving sixty-eight pages a copy. Despite printing fewer words, annual reports easily conveyed the main lessons of the war. According to the governors of the Queen's Hospital in 1918, the war above all demonstrated the need for more hospital beds, disabled soldiers alone requiring much treatment. Though more elaborate buildings were not deemed necessary, plans to relocate hospitals outside the city were commenced in these years.[112] All of these plans would be put into action following the cessation of hostilities.

[111] BCLLS, Queen's Hospital, Annual Report, 1918.
[112] Ibid.

CHAPTER 8

The University of Birmingham Medical School

I N 1900, the small provincial medical school originally run out of the Cox family private practice in Temple Row became the medical faculty of the new University of Birmingham. Despite affiliation with what was the first university campus in England, the new faculty, many of whom donned their new professorial titles, continued to occupy its old rooms at Mason College, only a short distance from its original premises in the centre of the city. While new buildings and modern facilities were being constructed for many of the university's scientific departments on 25 acres of land donated by Lord Calthorpe – 'on the Bournebrook side of the Edgbaston estate' – the medical school's links with voluntary hospitals firmly rooted the faculty in its city-centre site.[1] Unlike the science faculty, which moved 3 miles west of Chamberlain Square into its grand new buildings, medicine, like the majority of the city's voluntary hospitals, continued to occupy a visible place in the life of Birmingham's inhabitants. Symbolically, the school's location also emphasised the tensions that existed 'between those who upheld a vocational system of training and those who advocated a more academic, science-based medical curriculum as part of a university education'.[2] For the moment, the school's distance from the university continued to favour the former camp.

At the first recorded meeting of the university's medical faculty on 28 June 1900, the principal and newly promoted professors – those of Anatomy, Physiology, Medicine, Surgery, Midwifery, Gynaecology, Therapeutics, Forensic Medicine, Hygiene, Pathology, Lunacy, Operative Surgery and Ophthalmology – discussed only two issues: the colour of the academic gowns to be worn at degree congregations, exams and lectures, and the admission of women to the medical faculty.[3] In regards to the first subject, the faculty preferred an all black gown with cardinal watered silk. Despite staff debating the issue at some length, students began to disregard the rule of wearing gowns to exams and lectures as early as 1912. In general, it was unsuitable for a modern university, such as Birmingham, 'to impose the sight of such an anachronism as academic dress on the streets'. Unable to shed its industrial ethos, the city's university was destined to be set in its

[1] Ives, *The First Civic University*, 112.

[2] Waddington, *Medical Education at St Bartholomew's*, 146.

[3] UBSC, University of Birmingham, Faculty of Medicine, Minute Book, 1900–1903.

leafy suburbs. As a result, the undergraduate gowns and 'battered mortar-boards' of the town-dwelling medical students, according to the novelist Francis Brett Young, one of the university's famed alumni, lingered in a cloak room just inside the medical school.[4] Older customs, such as the presentation of academic medals and certificates of honour in medicine, also declined, though the school's five scholarships were more resilient and continued to attract medical students, including Young, to Birmingham, if only from the surrounding parishes.[5] By the outbreak of the First World War, all four of the school's main clinical prizes were awarded to women. Considered a more complex issue than academic dress, the instruction of women students continued to be discussed by the first medical faculty throughout the first summer of the twentieth century.

STUDENT AND EDUCATIONAL BODIES

Although women comprised 15 per cent of students at British universities in 1900,[6] their admission to English medical schools was a relatively new concept. A century earlier, European women had been professionally excluded from medicine, despite having provided the sick with care throughout history. The struggles of the first female medical practitioners, who travelled abroad to New York, Paris or Zurich from the mid-nineteenth century in order to qualify, have been regularly celebrated by historians.[7] Exploiting legal loopholes in order to obtain the qualification of the Society of Apothecaries in 1865, Elizabeth Garrett became the first woman to gain the right to practise medicine in England. Instrumental in the establishment of the London School of Medicine for Women in 1874, she both assisted and inspired many other women to follow in her footsteps, though less than half who commenced studies ever graduated in medicine.[8] The Medical Act of 1876 empowered all examining bodies, starting with the Irish College of Physicians – whose diploma was obtained by Sophia Jex-Blake, as well as Louisa Atkins[9] – to register the degrees

4 F. B. Young, *The Young Physician* (London, 1938), 233, 263.

5 The scholarships were the Ingleby, Sands Cox, Queen's, Sydenham and Russell Memorial Prizes.

6 C. Dyhouse, *No Distinction of Sex?: Women in British Universities, 1870–1939* (London, 1995), 7.

7 E. M. Bell, *Storming the Citadel: The Rise of the Woman Doctor* (London, 1953); C. Blake, *The Charge of the Parasols: Women's Entry to the Medical Profession* (London, 1990); T. N. Bonner, *To the Ends of the Earth: Women's Search for Education in Medicine* (Cambridge, MA, 1992).

8 S. Roberts, *Sophia Jex-Blake: A Woman Pioneer in Nineteenth Century Medical Reform* (London, 1993), 99.

9 *Lancet*, 1 November 1924.

of practitioners regardless of gender. The following year, the Royal Free Hospital, the first teaching hospital to admit women, was persuaded to open its wards to female medical students, though resistance never waned. The Royal College of Surgeons of England, for example, only accorded recognition to the London School of Medicine for Women and the Royal Free Hospital in 1909.[10] Those schools that admitted women to their lectures often found themselves the butt of rival institutions for allowing girls 'to share their masculine privileges'.[11] Additional hurdles, based on claims of limited accommodation or even declining institutional vitality, continually emerged. As a result, a number of institutions subsequently barred their admission or introduced quotas, claiming women were physically and mentally ill-suited to medicine, or ill-adapted to the 'strain' of education.[12] Nevertheless, 750 women had qualified by 1906.[13] While seven of London's dozen medical schools agreed to take female students during the First World War, only to ban them again during peacetime, none of the charters of the new provincial universities, including Birmingham, excluded women.[14] Their admission to provincial schools, however, had always been subject to strict conditions.

From October 1900 women began to attend classes at Birmingham's medical school, though not all areas of instruction were immediately opened to female students. Those classes not yet offered to women became accessible the following session, allowing the professors of these subjects time 'to arrange for the partial or entire separation of the sexes under [their] instruction'.[15] In the first weeks following the admission of women, the medical faculty appointed a qualified female demonstrator of anatomy, Violet Coghill, who went on to become house physician at the Queen's Hospital in 1916.[16] Coghill's classes took place in separate rooms, the female demonstrator exercising 'a general supervision over the women students'.[17] Additional female assistants were appointed to teach medicine, surgery, midwifery and gynaecology, and another was requested to assist the chair of forensic medicine, as some issues, such as rape, were deemed too sensitive to discuss before a mixed group of young men and women. In

[10] Cope, *Royal College of Surgeons*, 127.

[11] J. Manton, *Elizabeth Garrett Anderson* (London, 1965), 104.

[12] Waddington, *Medical Education at St Bartholomew's*, 296.

[13] Cope, *Royal College of Surgeons*, 124.

[14] Dyhouse, *No Distinction of Sex?*, 12.

[15] UBSC, University of Birmingham, Faculty of Medicine, Minute Book, 1900–1903.

[16] Queen's Hospital, General Committee Minute Book, 1907–17, HC/QU/1/1/7.

[17] Ibid.

a little over a decade, many of these female students, like Coghill, would actually deal with such serious matters in Birmingham's voluntary hospitals.

The faculty also began to discuss the status of clinical staff at the voluntary hospitals that sought official recognition by the university. Certificates of instruction from the General and Queen's were accepted, as were those of two municipal institutions, the Birmingham City Fever Hospital and Birmingham City Asylum. While the Queen's Hospital was additionally recognised for instruction in obstetrics, so were the Rotunda and Coombe hospitals in Dublin. The specialty of ophthalmology was also taught at the Queen's eye department, as well as the city's Eye Hospital. Associate status was similarly granted to the Royal Orthopaedic and Spinal Hospital and the Birmingham and Midland Ear and Throat Hospital. By January 1901, members of the clinical staff at both of the main general hospitals were also made university lecturers. A staff member at each hospital was appointed examiner, a post that was held for three years. All holders were paid as professors for their services.[18] The following month, both hospitals opened clinical instruction to women.

In 1900, undergraduates at the university totalled 189, with thirty-six enrolling at the medical school; only three, one from Hereford, and two from Carmarthen, came from outside the Birmingham region. This excluded part-time students, whose numbers swelled due to the popularity of postgraduate classes in the department of pathology and bacteriology. In total, the school had ninety-two students on its books. As the faculty's first meeting suggests, three of its three dozen new entrants were women (Edith Dora Grove of Walsall, Helen Gertrude Greener of Erdington and Florence Margaret Price of Carmarthen).[19] Along with Price, only two other students were the offspring of physicians, though medical students, including females, generally came from more affluent backgrounds than other university students.[20] In age, students ranged from sixteen to twenty-six years. As it was common for students to spend between six and eight years to complete their degrees, many of the youngest remained at the school until their mid-twenties. Interestingly, at the time of registration in 1900, the oldest student was a woman.

As this suggests, the admission of women greatly affected the culture of medical schools. In general, female students were older than their male colleagues, a fact that eased their transition into a number of medical schools.[21]

[18] UBSC, University of Birmingham, Faculty of Medicine, Minute Book, 1900–1903.
[19] UBSC, Dean's Register of Students, 1892–1929.
[20] Dyhouse, *No Distinction of Sex?*, 25.
[21] Waddington, *Medical Education at St Bartholomew's*, 310.

It also initially reduced the chances of any romantic liaisons between students, if the medical school environment had not already extinguished such possibilities. Ordinarily, the study of medicine had the potential to transform the way in which students thought about sex. At the very least, '[i]t knocks all the mystery out of [it]'; shows you exactly where you are 'instead of letting you go fumbling about in the dark'.[22] However, according to Francis Brett Young, '[e]ven if [female medical students] had not insulated themselves with shapeless djibbehs of russet brown, and bunched back their hair in a manner ruthlessly unfeminine, the common study of a subject so grossly material as anatomy would have rubbed the bloom from any budding romance.'[23] In the following years, medical education became even more segregated, the medical faculty creating separate ward classes for female students in 1905. In general, '[i]t was felt that in many cases full explanations of certain conditions were withheld by Teachers out of respect for the sex of the women students present'.[24] A year later, arrangements appear to have been successfully implemented, and the faculty reported that 'no practical difficulty appears at present to have arisen in respect of the existing arrangements at the two Hospitals for the teaching of Women Students'.[25] Though historians of the university may suggest otherwise, the continued existence of separate classes for women appears to suggest that a provincial medical education was very much characterised by a distinction of sex.[26]

TWENTIETH-CENTURY MEDICINE:
UPGRADING MASON'S COLLEGE

Like the governors of the city's hospitals, the faculty was far more concerned with upgrading available facilities than building new ones in these years. While scientific departments were occupying newly constructed premises in Edgbaston, the medical school continued to occupy buildings in the centre of Birmingham. Occupying premises built in the 1870s, many medical departments were increasingly regarded as run down and in need of modernisation. The physiology department, for example, was deemed 'unworthy of a University College to say nothing of a University'.[27] With little space, it was ill-fitted for three-fourths of the students who attended

[22] Young, *The Young Physician*, 296.

[23] Ibid., 268.

[24] UBSC, University of Birmingham, Faculty of Medicine, Minute Book, 1903–6.

[25] Ibid.

[26] Dyhouse, *No Distinction of Sex?*, 238.

[27] UBSC, University of Birmingham, Faculty of Medicine, Minute Book, 1900–1903.

its courses. No additional students could be admitted without duplicating lectures and demonstrations, a practice which was considered 'nothing short of an injustice'.[28] Some departments were moved to new premises, while others, such as therapeutics, still waited for their own laboratories, or even general furnishings as in the case of pathology. While no better off in terms of space, the Professor of Ophthalmology, Priestley Smith, 'occupie[d] the invidious position of being the only Professor giving a compulsory course [of twelve lectures] for which he [was] not paid'.[29] Under existing regulations he would not receive fees for another four years, which led the faculty in this case to award him a ten-guinea honorarium. While Smith did not receive payment for his hospital work either, his premises at the Queen's Hospital had recently been upgraded and rebuilt.

Other departments were equally in need of investment. Additional accommodation for the school's pathological collection was required, and the anatomical museum was considered as too small, with important recent additions, including the Simons-Freer collection, still packed away in boxes. The committee of the pathology museum was subsequently awarded a yearly grant of £50 to maintain the collection, preservative fluids costing at least £20, glass and other exhibition jars incurring another £30 expenditure annually. The Professor of Pathology continued to meet these expenses out of the department grant, a circumstance that was considered intolerable given that the museum was 'not a Department, but a University interest'.[30] As an indication of the museum's importance to instruction, three honorary assistant curators were appointed to assist the curator, Professor R. F. C. Leith. All other professors were called upon to help eliminate duplicates and deficiencies in the collection, and an architect advised on doubling the museum's size. Deficiencies in the library were to be rectified by raising its grant from £25 annually, which had permitted each member of staff to purchase a single new book each academic year. Meanwhile, therapeutics, forensic medicine and public health continued to wait for their own museums.[31]

As for other amenities and more 'modern' methods of instruction, an extension was required to accommodate a large projection microscope, the present room being too small and noisy. The faculty could hardly envisage the day, 'when engines and boilers and the annoying vibrations which

---

[28] Ibid.

[29] Ibid.

[30] Ibid.

[31] For more on medical museums in these years, see J. Reinarz, 'The age of museum medicine: the rise and fall of Birmingham Medical School's museum', *Social History of Medicine* 18, 3 (2005), 419–37.

they produce', no longer disrupted instruction.[32] Further down the list of priorities was the need to improve the student common room, which was described as 'a few removes better than a beer cellar'.[33] In his novels, Francis Brett Young, remembered it as a dismal chamber in the basement of Mason's College where students drank tea and ate 'squashed-fly biscuits'.[34] Outside meal times, the room attracted primarily unmotivated students, who spent most of their time lounging in basket-chairs playing the occasional game of poker and waiting for the racing results in the evening papers.[35] Like the common room, the college's other 100 rooms were 'low and ill-ventilated'. Besides better ventilation and additional space, the dissecting room required electric light because 'the drying nature of gas ha[d] a bad effect upon the [body] parts' which students dissected.[36] Aware of improvements at other medical schools, the faculty believed their institution was falling behind its provincial competitors. In particular, this led the dean to plead for more funds in 1901, as 'a policy of starvation may cripple this school at a time when its chances are the fairest'.[37]

The following spring, some progress was reported, beginning with the enlargement of the Department of Physiology. By October 1902, additional grants were made to the departments of Pathology and Therapeutics. A room in the former department was also set aside for histological work, while an entire lower laboratory was to be made available for the purposes of experimental physiology. In contrast to the rest of the building, the new histological room was described as large and well lit, the entire medical side of the building having been wired for electric lighting during the previous summer vacation.[38]

An application from the university senate for a vivisection license, which was granted in 1903, appears to mark the beginnings of research at the university and encouraged 'the development of a new type of teaching space – the teaching laboratory'.[39] Early histological experiments at Birmingham were concerned with the role of bacteria in digestion and with tuberculosis and its early cellular changes (in guinea pigs and rabbits). As in the 1870s,

---

[32] UBSC, University of Birmingham, Faculty of Medicine, Minute Book, 1900–1903.

[33] Ibid.

[34] Young, *The Young Physician*, 265.

[35] Ibid., 279.

[36] UBSC, University of Birmingham, Faculty of Medicine, Minute Book, 1900–1903.

[37] Ibid.

[38] Ibid.

[39] Waddington, *Medical Education at St Bartholomew's*, 116.

the laboratory's work also quickly encountered resistance. Unfortunately for the medical faculty, these years also witnessed an increase in sensitivity towards animal research following the publication of Louise Lind-af-Hageby's *The Shambles of Science* (1903). In particular, the Swedish anti-vivisectionist claimed medical instructors at University College, London, including the physiologist and discoverer of hormones William Bayliss, had tortured a brown terrier and 232 other insufficiently anaesthetised dogs in a number of animal experiments at the medical school which she attended in 1902.[40] Informed of the Birmingham faculty's application to undertake vivisection, animal rights campaigners delivered a petition to the university's authorities opposing similar experiments in the Midlands. In an effort to calm opposition, the chancellor of the university replied with a statement outlining his approval of scientific research, as well as his opposition 'to the torture of animals'.[41] He also reiterated that animals used in experiments would be anaesthetised, as required by the Cruelty to Animals Act (1876).

Other developments were less controversial and involved upgrading traditional forms of teaching space. Having recently refurbished their pathology museum, the faculty reorganised the anatomy museum in 1902 and established an ethnological museum for the use of the anatomy department. So popular were these museums that students immediately formed an Anatomical and Anthropological Society, and the dean of the medical school, Bertrand Windle, arranged for selections from its proceedings to be published in the *Journal of Anatomy and Physiology*.[42] Interest in anthropological subjects remained strong across the medical faculty and the university was identified in 1913 by Sir Richard Temple, the former Governor of Bengal, as one to which a school of anthropology was particularly suited.[43] Not only did the university already possess a museum and library to support such studies, but a greater knowledge of man he said was particularly relevant to the work of colonial officers.[44] Unfortunately, power struggles between departments unwilling to part with valuable collections stifled this particular initiative.[45] In contrast, extramural activities further developed with the establishment of a Medical Students' Literary

[40] C. Lansbury, *The Old Brown Dog: Women, Workers and Antivivisection in Edwardian England* (Madison, WI, 1985).

[41] UBSC, University of Birmingham, Faculty of Medicine, Minute Book, 1903–6.

[42] UBSC, University of Birmingham, Faculty of Medicine, Minute Book, 1900–3.

[43] *Birmingham Daily Post*, 28 October 1913.

[44] *Birmingham Daily Mail*, 7 November 1913.

[45] P. Sillitoe, 'Making links, opening out: anthropology and the British Association for the Advancement of Science', *Anthropology Today* 20, 6 (2004), 10–15.

Society, which published all nine papers presented by its members during its first session. In all likelihood, Francis Brett Young was a member of the society, if not its founder. According to *The Young Physician*, a novel which drew heavily on his experiences at Birmingham medical school, the society's members met in the smoking room next to the dissecting room.[46] What Young described as a 'smoking room' was in fact the new medical common room, on the Great Charles Street side of the building, opened in 1903.

EDUCATIONAL DEMANDS:
THE LIFE OF MEDICAL STUDENTS

Among other things, increased contact among medical students, whether in early literary or anthropological societies, enhanced their influence over medical education. Known to have frequently disrupted lectures and caused mayhem at the metropolitan medical schools, students exercised considerable control over the evolution of instruction in these years.[47] At Birmingham this was evident in a series of petitions delivered by students to the faculty in December 1902. According to these statements, which were signed by twenty students, the majority of discontent, as in Cox's time, related to clinical instruction. Given that the petitions, unlike faculty minutes, provide considerable detail concerning the students' workloads, as well as teaching standards, they are worth summarising briefly.

In general, students complained of apathetic instructors. For example, students attached to the Queen's Hospital claimed that surgical dressing was not sufficiently controlled and supervised by the officers, 'who do not show sufficient interest in the work of their dressers'.[48] Due to the limited time available to view casualties, it was suggested the Resident Dresser's period of office be shortened (to one month or six weeks), thereby permitting more students to occupy the post. Finally, students claimed that clinical lectures were not regularly delivered, even when advertised in the syllabus.

Students similarly criticised clinical instruction at the General Hospital. Those undertaking clerkships at the hospital desired fewer cases and requested these be discussed systematically by the honorary officers. Neither did the time of post-mortem examinations suit fourth-year students, who demanded that another dressership be instituted. Students also

---

[46] Young, *The Young Physician*, 305.

[47] K. Waddington, 'Mayhem and medical students: image, conduct, and control in the Victorian and Edwardian London teaching hospital', *Social History of Medicine* 15, 1 (2002), 45–64.

[48] UBSC, University of Birmingham, Faculty of Medicine, Minute Book, 1900–1903.

requested that 'Junior Medical Tutorials' be introduced at both main teaching hospitals and for longer periods (six months). The clinical committee similarly recognised the need for such classes, given a tendency among the senior teachers to 'teach over the heads of students', though also noting a deficiency in the elementary knowledge of students.[49] In future, lecturers were to be given full powers to select any cases for the purposes of demonstrations. Students further demanded separate senior and junior ward and clinical classes in medicine and surgery, which were to be kept distinct. While final-year 'men' were said to benefit from attending outpatient classes, junior students only hindered the classes without obtaining any corresponding advantage. Surgical dressing and clinical clerking additionally interfered with students' attendance at clinical classes and demonstrations.

A final petition complained of the amount of work students contended with while attending hospital. Most could not keep up with their reading and additional time would encourage better answers in exams and end cramming. For example, for third-year students in the winter term the day commenced only after hospital work concluded. On the completion of lectures most went home to read up on those cases observed during the day, and prepare for their exams in medicine, surgery, pathology and public health, as well as materia medica for the third professional exam. In the fourth year, students read for eight exams, two in medicine, surgery, midwifery and therapeutics. During the summer, students undertook postmortem work while preparing for exams in gynaecology, forensic medicine, toxicology and mental diseases, as well as the fourth professional exam, which consisted of pathology with practical work, public health, forensic medicine and toxicology. However, those students not passing their exams were not permitted to sit the professionals even though they would not take place for another two years. Consequently, students had to devote their main attention to examinations at precisely the time when they should have been concentrating on their clinical work, 'a fact which has been lately demonstrated at the Conjoint Exams' of the royal colleges of surgeons and physicians.[50]

Unlike in the first half of the nineteenth-century, when medical studies were largely self-directed, instruction had become highly structured. As a result, students' petitions may be interpreted as a response to the increasing regulation of medical studies. Despite existing dissatisfaction, student

---

[49] BCLA, General Hospital, Clinical Teaching Committee, Minute Book, 1892–1924, GHB 82.

[50] UBSC, University of Birmingham, Faculty of Medicine, Minute Book, 1900–1903.

numbers, with only minor exceptions, gradually increased in these decades. In 1901, thirty-one new students, including another woman, registered at the school.[51] Total student numbers were more impressive, with 192 registered in that year. In 1902, intake rose to forty-eight, though this was due in part to the admission of ladies preparing to be physical training teachers to the anatomical department, as well as a number of mature men preparing for their diplomas in public health. Only one to two dozen students in these years actually enrolled in the full medical course. The following year, new admissions declined to thirty-two and climbed only slowly, reaching forty again by 1907. Dividing students between the two main teaching hospitals appeared to reduce their numbers even more in the eyes of the clinical staff. Most students attended only one of the two main teaching hospitals for a period of three years. Few even attended the practice of the other associated hospitals or did so for only limited periods of time. Very few students attended the smallest hospitals, such as the Ear Hospital, whose board raised the cost of admission in order to attract students as clinical assistants, thereby avoiding fees, and remaining at the hospital longer. Clearly it was not just the students who required some time to adjust to the medical faculty's strict regulations. Less willing to compromise with teaching hospitals, the faculty threatened to remove the ear institution from its list of associated hospitals on this occasion and the hospital quickly rescinded its decision.[52]

The brief decline in student numbers between 1903 and 1907, though witnessed at other provincial schools,[53] was attributed to regional economic conditions. For example, as the medical school was located in an industrial region, many more opportunities existed for young local men to pursue careers in manufacturing and trade. Others attributed the decline to the school's entrance requirements, which had become too difficult. Additionally, the period of study was deemed too long, and the rewards comparatively small to justify tuition fees that had reached £184 for five year's instruction. Equally convinced of the first rules of economics, the faculty believed that the laws of supply and demand would eventually right the situation.[54]

[51] UBSC, Dean's Register of Students, 1892–1929.

[52] UBSC, University of Birmingham, Faculty of Medicine, Minute Book, 1900–1903.

[53] G. Grey Turner and W. D. Arnison, *The Newcastle upon Tyne School of Medicine, 1834–1934* (Newcastle upon Tyne, 1934), 217.

[54] UBSC, University of Birmingham, Faculty of Medicine, Minute Book, 1900–1903.

BEYOND THE LOCAL:
THE MEDICAL SCHOOL AND THE WIDER WORLD

While reflecting both the aspirations of inhabitants and the economic realities of the industrial Midlands, the school's horizons broadened somewhat in these years with the introduction of travelling scholarships and the arrival of foreign students. Established in 1902 by Mr George Myers in memory of his son, Dr Walter Myers, who died of yellow fever while conducting research for the Liverpool School of Tropical Medicine, the Myers Prize permitted students to combine their Birmingham studies with time abroad.[55] Whilst a previous generation had looked to France for enlightenment, medical educators from the middle decades of the nineteenth century grew more interested in the German schools. This was reflected in the regulations of the Myers award, which encouraged recipients of the £150 prize to attend a German university after obtaining a degree from Birmingham. Qualifying studies were to be in the area of pathology or clinical medicine, combined with pathological research, and the award could be renewed for a second year at a reduced stipend. The first recipient, John Dale, travelled to Hamburg in 1909 to pursue public health research.[56] Three years later, he returned to Birmingham and was appointed assistant medical officer of health to the city. Like postgraduate students, members of faculty also spent more time abroad as professional bodies developed international branches. Robert Saundby, for example, represented the faculty in 1903 at the fourteenth International Medical Congress in Madrid, where Ivan Pavlov read his paper on 'The Experimental Psychology and Psychopathology of Animals'.[57] Equally, while a previous generation had looked to other provincial hospitals and medical schools when undertaking reconstruction, members of the Birmingham faculty frequently prepared for redevelopment by conducting investigations abroad.

Many others looked to England for improvement, as is indicated by the number of foreign students who began to arrive in Birmingham in greater numbers at this time. The first student from abroad to register at the school was Tawfik Wasfi, a twenty-four-year-old Egyptian man, who enrolled in 1905 and obtained a teaching post at the Syrian Protestant College of Beirut after qualification.[58] The following year, two Indian students arrived, and many more Egyptian and Indian students enrolled over the following decade. By 1911, the school also began to attract students from China.

---

[55] Ibid.

[56] UBSC, University of Birmingham, Faculty of Medicine, Minute Book, 1906–11.

[57] UBSC, University of Birmingham, Faculty of Medicine, Minute Book, 1900–3.

[58] UBSC, Dean's Register of Students, 1892–1929.

Unlike former students, their arrival provoked criticism from local rate-payers, who were opposed to educating 'aliens and foreigners like Japs and Chinese'.[59] While some understood the benefits of training local students alongside those from other countries, subsequent waves of foreign students were targeted more carefully through advertising campaigns.[60] Consequently, in 1914, students, primarily the children of expatriates, began to arrive from South Africa and Canada. Most were already qualified and simply eager to supplement their medical knowledge through a period of postgraduate study.

While the medical faculty improved contact with practitioners working abroad, local links languished. For example, members of staff regretted 'that there [was] not a greater amount of touch between the Hospitals and the University'.[61] Contacts between the medical school and the specialist hospitals, forged only a decade earlier, were curtailed with the establishment of specialist clinics at the two main teaching hospitals. In some cases, few desired closer contact until institutions were improved. For example, with the 'exception of the conservation room', the Dental Hospital was not regarded suitable for 'special hospital work'.[62] In particular, the mechanical work room on the ground floor was 'hot and stuffy', given that it possessed no means of ventilation, besides a window (which was closed at the time of inspection) and a chimney. Occasionally, three gas-heated furnaces were used simultaneously in the room; as a result, the walls and ceiling were 'very dirty, covered, in fact, with soot'.[63] Though containing a double row of chairs for dental operations, the upstairs conservation room was 'insufficiently lighted for good work to be done'. A single water-closet for the use of patients of both sexes was equally inadequate; the students, numbering perhaps ten in a good year, used a water-closet outside the building.

As a result of such deficiencies, it was suggested that a new dental hospital replace the old building in no less than eighteen months. Although dental students numbered fifty-eight in October 1904, the hospital was described as 'the least satisfactory and most trouble-some institution connected with this University'.[64] With the medical faculty in Manchester about to institute a scheme for granting degrees and diplomas in dental surgery, it was decided to expedite construction in order to protect the interests of the dental department. Having previously conferred only

[59] *Birmingham Gazette*, 31 October 1911.

[60] *Johannesburg Star*, 8 June 1914; 6 July 1914.

[61] UBSC, University of Birmingham, Faculty of Medicine, Minute Book, 1903–6.

[62] Ibid.

[63] Ibid.

[64] Ibid..

degrees in dental surgery, the university also created a Licentiateship in Dental Surgery. By 1905, circumstances had drastically altered. With the enrolment of dental students increasing, the Birmingham faculty expressed the desire that relations between the university and Dental Hospital 'always be as cordial as now'.[65] They also recognised that dental students had not always 'received the consideration due to them, and in some measure the welfare of the dental student ha[d] been sacrificed to the requirements of the medical student'.[66]

PAEDIATRICS AND PATHOLOGY:
THE STRENGTHENING SPECIALTIES

Links with other specialist hospitals also improved with locally inspired changes to the curriculum. For example, although affiliated with the medical school in the nineteenth century, the Children's Hospital was designated an associated hospital of the university only in 1906. However, by the outbreak of war in 1914, the university still did not offer a course in the diseases of children, even though children were treated at both of the main teaching hospitals and the local maternity hospital. Apart from a lectureship instituted at Manchester's Victoria University in 1880, university appointments in the diseases of children were uncommon in nineteenth-century England.[67] The same lack of commitment prevented the emergence of a paediatrics journal in the country until 1896, decades after similar periodicals first appeared in German-speaking nations.[68] By the end of the century, a number of continental European universities began to appoint one or more chairs in paediatrics, whereas none was created in Britain until after the First World War. Appointed honorary professor of diseases of children in 1906, George Frederick Still's post in London was a college, rather than a university, chair. Birmingham made a similar appointment nearly a decade later in 1915, when Leonard Parsons became the university's first 'lecturer in infant hygiene and diseases peculiar to children'.[69]

During his first year, the Kidderminster-born and Mason College-educated paediatrician delivered ten lectures to fifth-year students. These included topics which had concerned previous generations of child specialists, including the new born infant, the development of children, rheumatic infections (pneumococcal and tuberculous) and specific fevers and

[65] Ibid.

[66] Ibid.

[67] Lomax, *Small and Special*, 158.

[68] F. H. Garrison and A. F. Abt, *Abt-Garrison History of Paediatrics, with New Chapters on the History of Paediatrics in Recent Times* (London, 1965), 125–30.

[69] UBSC, University of Birmingham, Faculty of Medicine, Minute Book, 1911–16.

diseases of the alimentary, circulatory and respiratory systems.[70] They also incorporated a new field of knowledge which had opened novel areas of study. With the concept of vitamins (coined in 1912), a number of apparently infectious illnesses were transformed into specific dietary deficiencies, which Parsons addressed in lectures on infant feeding (natural and artificial), as well as a module on the diseases of nutrition. A lecture on congenital mental defects, on the other hand, highlighted a growing interest in inherited disorders across medical disciplines.

In contrast to the late introduction of paediatrics to the curriculum, obstetrics was one of the first specialist requirements at medical schools. Introduced following a decision by the General Medical Council in 1886 that all students spend three months attached to a lying-in ward, requirements hardly advanced over the next two decades.[71] Having investigated the practices of other medical schools in 1906, the Birmingham faculty recommended all students attend at least twenty cases of childbirth.[72] Prior to attending any deliveries, students were to present themselves at all relevant lectures, successfully conclude their dresserships and clerkships and attend a lying-in hospital for a month. While some dissent surrounded the recommendation that students give their 'undivided attention' to obstetric practice during this month – as this would interfere with their ability to attend other courses – not all instructors favoured students attending deliveries at a lying-in hospital. Given the size of such institutions, it was believed that the time between deliveries would result in much wasted time. Moreover, the faculty advocated the direct supervision of students, which was less likely at a small specialist institution. As a result, members suggested that students train with certain recognised general practitioners.

It was with some hesitancy that the faculty recognised the Birmingham and Midland Hospital for Women as an associated teaching hospital of the medical school in October 1907.[73] Associate status, however, was granted somewhat more easily given the institution's reconstruction in 1905. By 1911, existing chairs of operative surgery and midwifery and diseases of women were nevertheless reduced to lectureships. From 1913, medical students were admitted to the Maternity Hospital, but only after they had commenced their fourth year of study. Students were required to complete twelve months of lectures before being admitted to the hospital's wards.

[70] UBSC, University of Birmingham, Faculty of Medicine, Faculty Calendar, 1916.

[71] R. Stevens, *Medical Practice in Modern England: The Impact of Specialisation on Medical Practice* (New Haven, CT, 1966), 43–44.

[72] UBSC, University of Birmingham, Faculty of Medicine, Minute Book, 1903–6.

[73] UBSC, University of Birmingham, Faculty of Medicine, Minute Book, 1906–11.

Their numbers were also restricted, six students being admitted to the wards at a time, their every move having been carefully monitored by the matron. Students attended the institution for one month, attending five lectures (or demonstrations) a week and six deliveries under the supervision of the house surgeon.[74] For this privilege each student, just as they had when the hospital opened in 1871, paid £2 2s.[75] As in the days of Lawson Tait, hygiene standards were rigidly maintained, all students adhering to methods for securing asepsis as laid down by the charity's Medical Board.

In contrast, the work of pathologists had evolved greatly since the 1870s. While the gradual introduction of the microscope and simple chemical tests on body fluids led to significant advances in diagnosis during this decade, the apparatus required was not extensive and the techniques simple enough to be used by clinicians themselves. Besides carrying out all post-mortems, pathologists undertook a range of hospital duties, from assisting in outpatient departments to administering anaesthetics at operations.[76] They also undertook teaching duties, spending much time with students, demonstrating morbid parts and instructing in microscopy. As a result, many of the first hospital pathologists were junior members of staff who gradually worked their way up to consultant status. For example, both Robert Saundby and Arthur Foxwell were for a time in the 1880s pathologists to the Women's Hospital before obtaining more senior posts and ultimately becoming consultant physicians. Professorships in pathology, though rare before the First World War, were similarly looked upon as stepping stones to honorary appointments as physicians and surgeons.[77] With the isolation of the tubercle bacillus in 1882, the cholera vibrio in 1884 and the pneumococcus and diphtheria bacillus in 1886, the practically useful science of bacteriology began to constitute the bulk of the pathologist's work in the last decade of the nineteenth century. Though slow to appear in England, facilities for diagnostic bacteriology were built in the late-1880s and 1890s. In 1899, when Otto Kauffman resigned as Professor of Pathology at Mason College, he was replaced by R. F. C. Leith. The following year, when Leith commenced his pathology course at the newly founded University of Birmingham, it comprised two parts, a systematic course of lectures on General and Special Pathology and one on Bacteriology with practical instruction in both subjects.

[74] UBSC, University of Birmingham, Faculty of Medicine, Minute Book, 1911–16.

[75] BCLA, Women's Hospital, Medical Board Minute Book, 1871–92, HC/WH/1/5/1.

[76] BCLA, General Hospital, Medical Committee Minute Book, 1868–76, HC/GHB/69; M. Weatherall, *Gentlemen, Scientists and Doctors: Medicine at Cambridge 1800–1940* (Woodbridge, 2000), 153.

[77] W. D. Foster, *Pathology as a Profession in Great Britain and the Early History of the Royal College of Pathologists* (London, 1982), 5, 15.

The medical school's facilities for undertaking bacteriological work greatly improved following Leith's appointment. Only a year after the establishment of the Pathological Society of Great Britain (1906), the university constructed a new Pathology building, which, on completion, was opened to the inspection of the medical officers of the district. While those visiting the department might have been impressed with its facilities, in many ways the work of the pathologist had hardly changed. While some academic pathologists in England deliberately refused any involvement in routine clinical 'services',[78] staff in the newly refurbished department spent much of their day carrying out routine tests and undertaking very little original research. While this might not have been tolerated in other departments, the ordinary work of the Pathology Department was supported because it generated an income for the medical school that could be used to expand the pre-clinical science facilities more generally.[79] The majority of funds derived from the examination of specimens for the city's public health department. For example, in 1910, the department's staff analysed forty-five samples of water and 448 of milk. They also examined tumours (126) and sputum (1,102), and tested for diphtheria (5,263) and typhoid (315). Along with 591 miscellaneous tests, the department undertook a total of 7,880 analyses.[80] Though the department's bacteriological work continued to grow over the next decade, Leith's chair in pathology and bacteriology was divided in two only in 1919, when he resigned his post;[81] a similar decision had been made in Manchester fifteen years earlier.[82] The commercial work of the department, on the other hand, was placed under a separate officer.

Though a division of labour considerably improved instruction in pathology, the distribution of university departments over two sites caused numerous difficulties throughout these years. When the departments of Chemistry and Physics moved to Bournbrook in February 1908, for example, more accessible classes for dental students, who remained based in the city centre, were organised. With Biology and Anatomy still at Edmund Street, changes to the timetable were introduced in 1910 in order that first-year medical students could spend every Monday and Wednesday and Thursday morning at Bournbrook. Tuesday, Friday and Thursday afternoon

[78] H. Valier, 'The Politics of Scientific Medicine in Manchester, *c.* 1900–1960' (PhD thesis, University of Manchester, 2002), 57.

[79] Sturdy and Cooter, 'Science, scientific management, and the transformation of medicine', 442.

[80] UBSC, University of Birmingham, Faculty of Medicine, Minute Book, 1906–11.

[81] UBSC, University of Birmingham, Faculty of Medicine, Minute Book, 1916–21; *Birmingham Post*, 17 December 1936.

[82] Valier, 'The Politics of Scientific Medicine', 78.

and Saturday morning, on the other hand, were spent at Edmund Street.[83] Though the location of the teaching hospitals would keep students in the city centre, further administrative changes brought the hospitals and the university into closer proximity.

RECOGNITION AND RESULTS:
CLINICAL BOARDS AND EXAMINATION BOARDS

In February 1911, with the reconstitution of the University Clinical Board, clinical lecturers became members of the university. A similar relationship with the dental school was established when its board was constituted in July 1912. Above all, this administrative change brought clinical instruction under the direct control of the university. Previously appointed by the two main teaching hospitals, the reconstituted board comprised nine members, four from the staff of the General and Queen's hospitals and five university employees.[84] Administrative changes also brought financial advantages. With university representation strengthened, medicine, like engineering, mining and metallurgy, became eligible for financial assistance from the newly created Board of Education (1899). Set up to supervise the educational provision of local authorities, the Board developed a financial interest in the organisation of medical education when it provided its first grant to St Mary's Hospital in 1908; a decade later, with St Mary's grant increasing to £3,000, the Board's funds were equivalent to half the sum obtained from student fees.[85] A poor financial climate encouraged additional metropolitan schools, as well as provincial ones, to apply for the Board's grants in order to meet the costs of teaching and research, as well as laboratories. Totalling approximately £1,000, Birmingham's first grant was received in July 1912.[86] In exchange, the Board began to exert influence on the way in which medical education was organised, insisting for example that the payment of clinical teachers be placed on a defined basis.[87] Half the sum subsequently went to the clinical instructors, who were paid £5 each for organising lectures, ward classes, demonstrations in outpatient departments, tutorials and pathological demonstrations.[88]

Even before the institution of these changes, clinical instruction was subjected to more rigorous terms and conditions. In November 1908, for example, the clinical board determined that professors who no longer

---

[83] UBSC, University of Birmingham, Faculty of Medicine, Minute Book, 1906–11.

[84] UBSC, University of Birmingham, Faculty of Medicine, Minute Book, 1911–16.

[85] Heaman, *St Mary's*, 148.

[86] UBSC, University of Birmingham, Clinical Board Minutes, 1911–26.

[87] Waddington, *Medical Education at St Bartholomew's*, 171.

[88] UBSC, University of Birmingham, Clinical Board Minutes, 1911–26.

provided clinical instruction at Birmingham's teaching hospitals were not eligible to be examiners in their areas of specialty.[89] The following March, a retirement age of sixty-five for lecturers and professors was also introduced. The first to retire under the new rules in June 1910 was the eminent ophthalmologist Priestley Smith. However, given his long association with the medical school and his international reputation, he was invited to hold the chair for an additional two years.[90]

In most cases, teaching posts and departments underwent their greatest reorganisation following a retirement. Significantly, after Priestley Smith's resignation, the professorship in ophthalmology was discontinued and transformed into a lectureship, though classes in the subject were increased from twelve to twenty-two. The first to occupy the position was J. Jameson Evans. Examinations in ophthalmology, formerly held at the Queen's Hospital, were held at the Eye Hospital from June 1913, as it was there that Jameson Evans was otherwise posted. The following winter the schedules of four students undertaking work at the hospital were accepted as clinical instruction in ophthalmology.[91] A subsequent proposal to make the Eye Hospital an associated hospital, however, was withdrawn in the summer of 1914. Additional specialist classes covered the subjects of ear and throat, skin and gynaecological cases. Those addressing the ear and throat were least well attended, at times attracting only a couple of students.[92] Valued at £10, the George Henry Marshall (Ophthalmic) Scholarship, generated some additional interest in ophthalmology after 1907. In general, though represented by their own hospitals in the city, specialties occupied little space in the medical school curriculum during these years.

Given a lack of evidence, instruction in most subject areas is perhaps best judged by the performance of students in examinations. The percentage of Birmingham students rejected by licensing bodies in these years remained high, more than 50 per cent failing their examinations in 1910 alone. Comparatively, failure rates of students in Manchester for these years ranged between 23.6 and 41 per cent, while 23 per cent of students at Leeds failed in all categories (medicine, surgery and midwifery).[93] Results for dental surgery were much better. Between 1904 and 1911, the average rejection rate of all Birmingham students had been 39 per cent, though results for 1909 were also particularly poor; of nineteen students, only nine

[89] UBSC, University of Birmingham, Faculty of Medicine Minutebook, 1906–11.

[90] Ibid.

[91] UBSC, University of Birmingham Medical School, Minutes of the University Clinical Board, 1911–26.

[92] Ibid.

[93] UBSC, University of Birmingham, Faculty of Medicine, Minute Book, 1911–16.

passed. The results for 1911 were better, with eleven passes and five failures.[94] Despite improved examination results, according to Robert Saundby, '[i] lliteracy among medical students was appalling'. Should the final examination have been in composition and spelling instead of medicine, candidates would have been 'ploughed'.[95] Dental students, despite posting better results, were described by instructors at the General Hospital, whose classes they attended, as 'entirely ignorant of anatomy and physiology'. Their presence at some clinical classes, such as elementary medicine, 'only hamper[ed] the class for any useful purpose for the teaching of medical students'.[96] Such views went unchallenged by subsequent results. By 1914, 44 per cent of Birmingham's medical students continued to be rejected by the London licensing bodies. The same year, thirteen of seventeen (76.5 per cent) of those who attempted the final MB in dental surgery also failed.[97]

SCHOLARLY STRUGGLES:
MEDICAL EDUCATION AND THE WAR

As one might imagine, the outbreak of war considerably disrupted medical education throughout the country. By October, thirty-one Birmingham students, nearly as many as had enrolled at the medical school that year, entered military service. In order to provide greater numbers of qualified medical personnel, a special early final examination was instituted at the end of the academic year. The annual conversazione was the first of many ordinary academic events to be cancelled the following year, and attendance at all other events, including lectures, declined. As in the past, the dean addressed students on their poor attendance at lectures and classes. However, on this occasion it was discovered that those absent were holding resident appointments at the General Hospital owing to the present shortage of qualified men. Instead of being reprimanded, they were requested to make every effort to attend classes, and each case would be separately considered by the faculty at the end of the session.[98] By June 1915, when the first student death was reported, the faculty's stance on attendance, among other disciplinary issues, became even more tolerant. By war's end, R. W. L. Edginton, who had entered the medical faculty in October 1913 before being promoted to Second Lieutenant in the 5th Battalion Royal

---

[94] Ibid.

[95] *Birmingham Gazette*, 10 June 1912.

[96] BCLA, General Hospital, Medical Committee dealing with Clinical Teaching, 1892–1924, GHB 82.

[97] UBSC, University of Birmingham, Faculty of Medicine, Minute Book, 1911–16.

[98] Ibid.

Warwickshire Regiment, was only one of 175 university staff, students and alumni who lost their lives in battle.[99]

With so many young men enlisting for military service, the presence of women at the medical school was also felt more acutely in wartime. Besides increasing as a proportion of total medical students, female students were awarded all four main clinical prizes in 1914. In the summer of 1915, a local newspaper campaign further attempted to boost the enrolment of women in the city.[100] Of thirty-six new medical students enrolled in October 1915, eleven (31 per cent) were women.[101] Three months later, only women and men under military age were permitted to continue their studies. As a result, entry at the London schools, with their discriminatory entrance policies, was down 25–45 per cent, while enrolment at the London School of Medicine for Women reached 110, compared with an average of forty-four in the three years before the war.[102] By 1917, women comprised 40 per cent of medical students at Birmingham. So, too, was the presence of foreign students more noticeable. In January 1915, Leopold Leon Joseph Stappers, the first of many Belgian refugees, was admitted to the first-year medical course with remission of all fees on recommendation of the University Senate.[103] By June 1916, reciprocal agreements between English and Canadian medical schools were also introduced.

Like those of the city's teaching hospitals, the facilities of the university and medical school were also placed at the disposal of the military authorities during the hostilities. By the end of 1914, for example, the Aston Webb buildings had been transformed into the 800-bed First Southern General Hospital. Professor Leith's pathological and bacteriological department gave up much of its analytical work and was also placed at the service of the Southern's staff. During this period, laboratory workers systematically examined the discharges of all wounds, as well as all cases of tetanus and gangrene. They also undertook countless blood examinations, Wassermann reactions, throat swabs, tested sputum for TB, faeces and urine for specific germs and prepared and supplied vaccines, including that for typhoid. For the duration of the war, their workload only increased as more university buildings were turned over to the Southern, which, by the time the Armistice was signed in November 1918, had control of 3,293 beds.[104] While the faculty decided eight months earlier that the work undertaken

[99] *Birmingham Mail*, 8 June 1915; Ives, *The First Civic University*, 164.

[100] *Birmingham Daily Post*, 20 July 1915.

[101] UBSC, University of Birmingham, Faculty of Medicine, Minute Book, 1911–16.

[102] *British Medical Journal*, 30 October 1915.

[103] *Birmingham Daily Post*, 6 November 1914.

[104] Ives, *The First Civic University*, 166.

in military laboratories would count towards the lab requirement of the public health diploma,[105] evidence suggests most students carried out far more practical work during the war than was common in peacetime.

Like hospitals which found it increasingly difficult to procure supplies and attract medical officers, the medical school faced the additional difficulty of procuring examiners in these war years. They were, however, able to assist hospitals in overcoming some of their deficiencies. For example, with the backing of the War Office, fifth-year medical students were permitted to take the place of qualified residents at military hospitals. Unlike in peacetime, when medical students paid for clinical instruction, during the war, students as a result of such services were paid 3s 6d a day and received rations, accommodation and a half-a-crown-a-week washing allowance.[106] Those appointed to approximately a dozen paid posts locally were permitted time off most afternoons for lectures and attended fever and asylum courses on Saturday mornings. All work undertaken by students at the First Southern General Hospital was accepted in lieu of clinical instruction at the Queen's and General hospitals and the usual tutorial classes, though some experience at the two main teaching hospitals was regarded as beneficial before serving at military hospitals.[107]

POST-WAR PEDAGOGY:
CONTINUITY AND CHANGE IN PEACETIME

For the majority of senior male students in military service, the medical school changed considerably during their absence from Birmingham. To begin with, some subjects were no longer regarded as relevant and dropped from the curriculum, including Latin, which was removed in January 1916.[108] Psychology, though very relevant to medical studies, had become too large a subject to be included in the undergraduate curriculum. Already overloaded, the curriculum became even more congested with the introduction of a short course on radiography in 1918, which was taught by Harold Black, radiographer to the General Hospital. Instruction in preventive medicine, encouraged by the General Medical Council, was incorporated into existing courses. While some courses already considered (albeit briefly) the effects of environment and occupation on health, additional lectures on diet, exercise, Child Welfare Schemes, as well as the use of outpatient departments to teach preventive medicine were introduced.[109]

---

[105] UBSC, University of Birmingham, Faculty of Medicine, Minute Book, 1916–21.

[106] Ibid.

[107] Ibid.

[108] Ibid.

[109] Ibid.

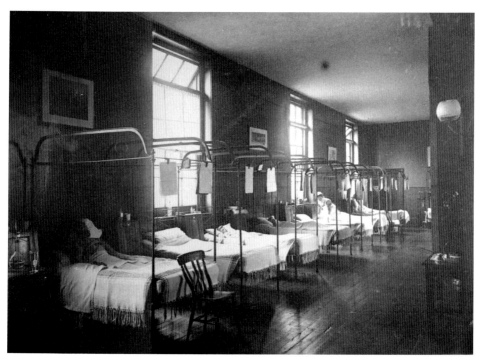

23 Ward 6, a men's medical ward, at the old General Hospital, *c.* 1894

24 Two nurses in the Large Operating Theatre, new General Hospital, *c.* 1898;
photograph by J. Hall Edwards, radiologist

42 Birmingham Hospitals Centre, 1938

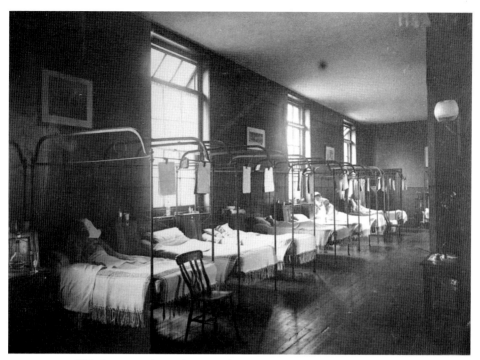

23 Ward 6, a men's medical ward, at the old General Hospital, *c.* 1894

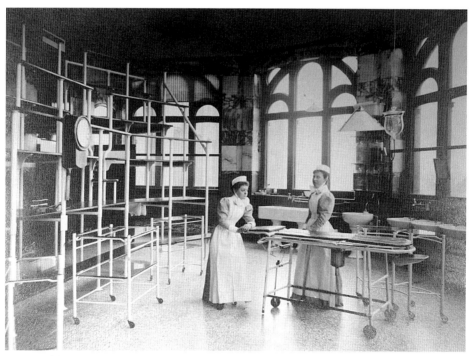

24 Two nurses in the Large Operating Theatre, new General Hospital, *c.* 1898;
photograph by J. Hall Edwards, radiologist

25 Before and after photographs
of a young female patient,
Orthopaedic Hospital, 1914-15

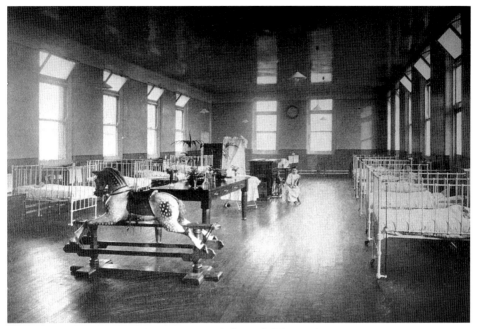

26 Children's ward, Queen's Hospital, *c.* 1910

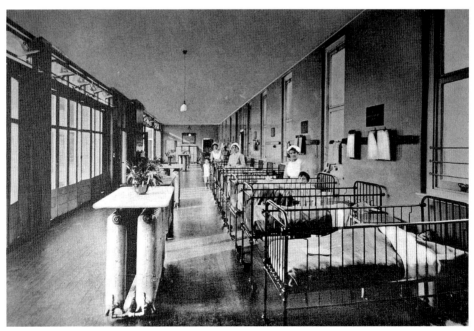

27 Ward in the new Children's Hospital, Ladywood Road, *c.* 1920

28 Nurses in the Conservatory at the General Hospital, *c.* 1900

29 Verandah of the new Margaret Ward, Women's Hospital, Showell Green Lane, 1913

30 East Filling Room, Dental Hospital, Great Charles Street, *c.* 1905

31 Female medical students, University of Birmingham medical school, *c.* 1915-16. Hilda Lloyd is sitting in the front, second from right.

32 Nursery at the Maternity Hospital, Loveday Street, 1933

33 Waiting room at Ear Hospital, Edmund Street, 1933

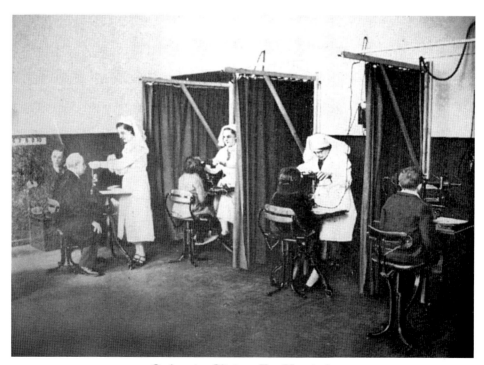

34 Orthoptics Clinic at Eye Hospital, 1937

35 Massage Department at the Woodlands branch of the Orthopaedic Hospital, 1934

36 Boot and splint-making shop at the Forelands,
Royal Orthopaedic Hospital, Birmingham, 1930

37 Operating Theatre at the Women's Hospital, Showell Green Lane, 1936

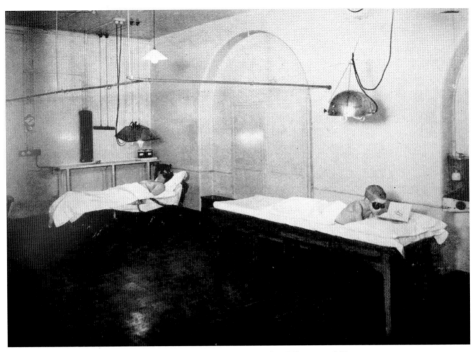

38 Patients undergoing treatment in the Electrical Department,
Skin Hospital, John Bright Street, 1934

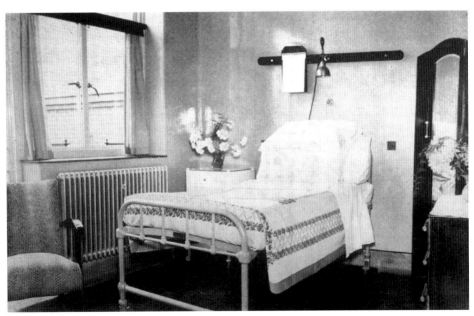

39 Private Radium Ward, Women's Hospital, 1934

40 Burcot Grange at Blackwell, Warwickshire served as a convalescent home to the Eye
Hospital from 1937

41 Architect's Drawing of Leonard Parson Block, Children's Hospital, *c.* 1940

42 Birmingham Hospitals Centre, 1938

Though a Department of Biochemistry with an associated chair was also recommended, in March 1920, an Assistant-Professor in Biochemistry was appointed instead to assist the Chair of Physiology.

Mirroring the expanding curriculum, nationally, student numbers dramatically increased after the cessation of hostilities with the introduction of government grants to ex-servicemen. By February 1919, urgent assistance was needed to teach anatomy at the school as student numbers had increased dramatically. The easiest way to create additional room was to set up separate classes for dental students, who had previously attended many classes alongside medical students. The repetition of classes was the only other option where space did not permit larger groups. Given the increase in numbers, roll call had become impractical and signing sheets were introduced into classes where attendance was greater than thirty. Peacetime also witnessed the return of pre-war sex ratios. Of the eighty-two students commencing their studies in October 1920, only eight were women.[110]

The increase in student numbers also extended to postgraduate students. Though rarely addressed by historians of medical education, postgraduates were a more noticeable constituency at British medical schools during the interwar period, although teaching often remained 'limited and ineffectual'.[111] Berlin and Vienna, which had attracted large numbers of students before the war, were no longer attractive centres for graduate work. As a result, research activity increased noticeably at a number of English medical schools which supplemented their postgraduate diploma courses in Public Health with Doctor of Philosophy degrees in order to encourage research.[112] Those undertaking postgraduate work were required to have a medical degree already and to have carried out laboratory work for at least two years. With many more students contemplating graduate work, staff could also envisage ways to lighten the undergraduate curriculum, especially the final year.

Postgraduate medical education was originally formalised in Birmingham with the formation of the Midland Medical Society in 1843. In 1875, the Birmingham Medical Institute was founded and, by the following year, boasted a membership of 172.[113] In 1880, the Institute further provided the Society, and its vast library, as well as two female members, with a home of its own when its original building in Edmund Street was

---

[110] Ibid.

[111] Waddington, *Medical Education at St Bartholomew's*, 334.

[112] UBSC, University of Birmingham, Medical Faculty Minutes, 1916–21.

[113] G. M. Goodman, *A Victorian Surgeon: A Biography of James Fitzjames Fraser West, 1833–83, Birmingham Surgeon* (Studley, 2007), 221.

opened.[114] Much of the society's postgraduate work, like that of the medical school, consisted in hosting lectures in fields of special interest. After 1900, along with established annual lectures, the faculty of medicine at the University of Birmingham introduced courses to improve links with local practitioners. A course in higher physics, for example, was offered in 1903 and covered Roentgen rays, radium and other matters of 'present interest'.[115] By 1920, the school, like many others,[116] offered short summer sessions in every medical specialty. Additionally, regional educational efforts had been consolidated into a national scheme for postgraduate study with central offices in London. Under the scheme, which had been advocated by William Osler since the end of the First World War, a register listing all postgraduate lectures, demonstrations and classes for practical work was established. A year later, a committee was set up under the chairmanship of the Earl of Athlone to establish a specific school for postgraduate medical education attached to a dedicated teaching hospital. When none of the existing teaching hospitals appeared interested in developing the scheme, it was decided to site the school in a London County Council institution, the Hammersmith Hospital. Formally opened in 1935, its full-time clinical professors were provided with the opportunity 'to blaze a perfectly new road, untrammelled by tradition, vested interests or medical dead wood'.[117] Besides a brilliant research record, the school convinced many more British doctors that medical education should be reorganised.[118]

By 1920, members of Birmingham's medical faculty regularly discussed the revision of undergraduate medical education. Nationally, debates were equally active, with that relating to the value of science to medical education growing more intense.[119] Others argued for a return to traditional notions of clinical medicine, concerns which were fuelled 'by the worry that a multiplicity of tests were being substituted for good clinical practice'.[120] In general, members of the Birmingham faculty put forward two standards of curricula, one lasting five years, the other six.[121] With the understanding

---

[114] G. W. Hearn, *Dudley Road Hospital, 1887–1987* (Birmingham, 1987), 104; Goodman, *A Victorian Surgeon*, 221.

[115] UBSC, University of Birmingham Medical Faculty Minutes, 1903–6.

[116] Waddington, *Medical Education at St Bartholomew's*, 334.

[117] C. C. Booth, 'Half a century of science and technology at Hammersmith', *British Medical Journal* 291 (1985), 1771–9.

[118] F. N. L. Poynter, 'Medical education in England since 1600', in *The History of Medical Education*, ed. C. D. O'Malley (Los Angeles, 1970), 235–50.

[119] Waddington, *Medical Education at St Bartholomew's*, 183.

[120] C. Lawrence, 'A tale of two sciences: bedside and bench in twentieth-century Britain', *Medical History* 43 (1999), 435.

[121] UBSC, University of Birmingham, Medical Faculty Minutes, 1916–21.

that at least half of their time be devoted to teaching, it also became clear that medical school staff required better pay if education was to improve. Laboratories and other research facilities also required development if the bench was ever to supplement a practitioner's bedside skills.[122] Finally, it was suggested that courses on children's diseases, the eye, mental health and public health be made compulsory. Afflictions of the skin, ear, nose and throat, venereal, radiography and electrical treatment, on the other hand, were not made compulsory. Even though this theoretically left the clinical material of two specialist hospitals outside the fold of the medical curriculum, it was advocated that all of the city's hospitals, not just the few, be tied to teaching.[123]

By the interwar period, the hospital's role in medical education was universally valued. According to the members of Birmingham's medical faculty, clinical teaching in hospital led the student to become a keen observer, fostered a 'close personal relationship' between student and patient and cultivated a sense of responsibility.[124] However, in their opinion, hospital work had also become too practical, leaving students insufficiently acquainted with the beginnings of disease, only its end results. Not only were more beds required in Birmingham, but clinical teaching in many respects remained unsystematic. Consequently, by the interwar years, members of the university's medical faculty spent less time discussing the place of women at medical school and devoted considerably more time to clinical instruction. Though full-time clinical units as established at the Hammersmith and in the United States remained contentious, their introduction at British medical schools had become inevitable. Combining laboratory and clinic, most agreed, would place medicine on an academic footing and 'extend the frontiers of knowledge'.[125] At the very least, it promised to be a useful corrective to the 'old empirical approach' of medical education.[126]

[122] *Birmingham Daily Post*, 10 February 1919.

[123] Ibid.

[124] UBSC, University of Birmingham, Medical Faculty Minutes, 1916–21.

[125] G. Newman, *Recent Advances in Medical Education in England: A Memorandum Addressed to the Minister of Health* (London, 1923), 89–91.

[126] Waddington, *Medical Education at St Bartholomew's*, 193.

CHAPTER 9

The Hospitals in the Interwar Period

THE DEFERRAL of much building work at Birmingham's teaching hospitals by the outbreak of war in 1914, not to mention increased wear and tear of premises during the conflict, ensured the interwar period would be characterised by reconstruction. Though the work of many hospitals during hostilities was subsidised by grants from the Home Office, such funds dried up after 1919 and did not permit the sort of rebuilding that had become obvious to many during the war effort.

Instead, governors aimed to reduce costs and avoid duplication between voluntary institutions by co-ordinating services more effectively. Paradoxically, efforts to co-ordinate hospital work in order to save money and prevent state intervention led to the creation of not one, but two regulatory bodies, the Joint Hospitals Council and the Local Voluntary Hospitals Committee, the latter having been created specifically to dispense £500,000 to voluntary hospitals following a report by the Cave Committee in 1921.[1] Duplication in the regulatory sphere ended in 1925, when these organisations joined forces to become the New Hospital Council, which aimed to maintain the voluntary system, co-ordinate appeals and purchases, and organise systematic contributions from employers and employees. While it has been suggested that the financial situation of provincial hospitals in these years was 'less acute than in London', others believed that the boards of the Birmingham voluntary hospitals needed to make greater efforts if they were to survive the interwar period.[2] Co-operative efforts culminated with the decision to amalgamate the General and Queen's hospitals in the mid-1930s and transfer their services to a new Hospitals Centre in Bournbrook.

By this time, modern hospital finance 'had outgrown the old ticket system'.[3] Unable to rely on traditional sources of income, such as subscriptions, governors turned their attention to a more reliable donor, the paying patient, who became 'the linchpin of hospital income' in these years.[4] Though greater proportions of hospital income derived from registration fees and private wards, the vast sums required for maintenance came from

---

[1] S. Barnes, *The Birmingham Hospitals Centre* (Birmingham, 1952), 36–7.

[2] M. Gorsky, J. Mohan and M. Powell, 'The financial health of voluntary hospitals in interwar Britain', *Economic History Review* 55, 3 (2002), 535.

[3] *Birmingham Daily Post*, 17 March 1927.

[4] Gorsky, Mohan and Powell, 'The financial health of voluntary hospitals', 547.

hospital contributory schemes.[5] Originating with the Hospital Saturday Fund, Birmingham's contributory scheme provided what many low-income families regarded as a cheap insurance against the risks of hospitalisation. In return, voluntary hospitals obtained the funds required to maintain and modernise facilities during the interwar period, collections in 1936 alone raising £381,000, a sum equivalent to 85 per cent of hospital costs.[6] Consequently, by this time many saw a future for voluntary hospitals beyond these years and even up to the formation of a National Health Service in 1948. The survival of a number of Birmingham's voluntary hospitals, however, was far less certain at the beginning of the interwar period.

ADENOIDS AND PRIVATE PATIENTS:
THE EAR HOSPITAL

Like Birmingham's other hospitals, the Ear Hospital commenced the interwar period with a deficit. At £657, however, it was manageable, but only if plans for expansion remained modest. Treating 1,942 inpatients and 10,254 new outpatients on an annual budget of £6,498 in 1920, subscriptions still accounted for half of income.[7] Other fiscal innovations lightened the financial burden only slightly. For example, an offer from the Cadbury family to refund the cost of treating their employees was rarely replicated by other employers.[8] As a result, the inauguration of a Linen League, though promising only minimal savings each year, was worth as much, if not more, in the long term as a grant (£550) received from the Ministry of Pensions for war work.

By 1921, post-war alterations commenced. With tonsillectomies in particular becoming fashionable and consequently increasing in number (1,229), governors purchased premises at Ludgate Hill for use as a tonsil and adenoid clinic. While reducing the hospital waiting list, the fourteen-bed ward increased costs by £3,000, half of which was covered by the sale of war stock and a £500 donation from Mitchells & Butlers brewery. Opening in May 1922 with a debt of £1,693, the ward treated 935 children in its first year, and regularly closed due to outbreaks of infection, as opposed to an inadequate income.[9] Though not yet described as an ENT hospital, the

[5] S. Cherry, 'Beyond National Health Insurance: the voluntary hospitals and hospital contributory schemes: a regional study' *Social History of Medicine* 5, 3 (1992), 455–82; M. Gorsky, J. Mohan and T. Willis, 'Hospital contributory schemes and the NHS debates, 1937–46: the rejection of social insurance in the British welfare state?' *Twentieth Century British History* 16, 2 (2005), 170–92.

[6] *Birmingham Daily Post*, 15 February 1937.

[7] BCLLS, Ear Infirmary, Annual Report, 1920.

[8] Ibid.

[9] BCLLS, Ear Infirmary, Annual Report, 1922.

work of the adenoid ward necessitated the appointment of a second house surgeon and bacteriologist, Dr A. F. Wright. Equally significant was the election in 1923 of Miss Agnes G. Brough, the institution's first female officer.

In 1930, premises on the Hagley Road in St Chad's, just west of the city centre, were acquired, treating almost as many patients as the Ear Hospital's main fifty-bed hospital, which admitted 1,060 that year. Paying up to 10s 6d a day for treatment, patients at St Chad's quickly attained privileged status. However, by 1932, charity patients were also admitted to the annex's day wards in order to reduce waiting times. Though not yet self-supporting, the hospital's income surpassed £9,000 in 1930. More importantly, subscriptions accounted for only £1,300 of this amount, two-thirds of the hospital's costs being met by the Birmingham Contributory Scheme. Though a new building was urgently needed, funds permitted the acquisition of a lift and new sterilising apparatus, both powered by electricity.[10] By relying on the Eye Hospital's radiography equipment, governors reduced their deficit to £43 by 1933, but continued thereby to live in the shadow of their early and more successful rival. Relations between many other hospitals, even those with no previous connections, would be characterised by similar co-operative efforts in these years.

In 1934, more extensive changes were initiated. With services at St Chad's handed over to Smethwick Corporation, additional accommodation for private patients was secured at Kingsthorpe Nursing Home in Edgbaston. The main hospital was to be extended on the site of Birmingham's old stock exchange, making room for its own x-ray department, while hay-fever and hearing-aid clinics, were also proposed; while the former clinic, reflected the emergence of both immunology and the laboratory in ENT work, with many allergy sufferers receiving subcutaneous injections to encourage tolerance to pollen, it also maintained a place for the more traditional approaches of the botanist and meteorologist.[11] The outpatient department gained a new casualty operating theatre, dispensary and registration office. With beds reduced to forty-five during reconstruction, premises in the Hagley Road, with room for seventeen patients, were used as a recovery home. Following extensions, the hospital had seventy beds, seventeen at the recovery home and eleven private rooms. The Ludgate Hill site, on the other hand, closed in 1936 and was converted into a home for seventeen nurses. Costing £15,000 and making the hospital the largest of its kind outside London, the extension was opened in

[10] BCLLS, Ear Hospital, Annual Report, 1930.

[11] M. Jackson, 'Allergy and history', *Studies in History and Philosophy of Biological and Biomedical Sciences* 34, 3 (2003), 395.

June 1938 by the author Francis Brett Young, a former medical student at Birmingham. Medical staff at the enlarged hospital consisted of three house surgeons, who treated 2,535 inpatients and another 8,739 outpatients (50 per cent of whom were children).[12] While dental students had been attending its wards the previous six months, governors urged students to take greater advantage of the hospital's clinical cases; it is unlikely that Young ever did.

ORAL EXAMS:
THE DENTAL HOSPITAL

Students more regularly attended the Dental Hospital during this period. In 1920, forty-three new students entered its school, fifteen of whom joined the mechanical department where they learned the art of denture-making. In total, the hospital recruited seventy-nine dental students and forty-three mechanical pupils, many in receipt of government grants. During their studies, students progressed through five main departments – extraction, anaesthetic, prosthetic, orthodontic and conservancy – the majority of time being spent in the last division. In the conservation room, students treated 4,860 patients and performed 42,000 operations during 21,940 attendances in 1920.[13] Though numerous, other dental hospitals were equally busy in the days following the passage of the Dental Act (1921), which, much like the Medical Act (1858), aimed to prevent the 'intolerable' activities of the unqualified practitioner.[14] Previously excluded from practice, women were encouraged to study dentistry, though few immediately enrolled.[15] The first qualified female practitioner at the hospital, Dr Doris Willcox, was appointed as a stipendiary anaesthetist in 1922.

Besides a complete radiographic department, donated by the University Dental Clinical Board, the immediate post-war years at the Dental Hospital were characterised by redecoration, not reconstruction. As a result, the charity's deficit declined from £1,000 (1920) to £181 in four years. Advertising costs (£2 10s) were also finally under control, the majority being devoted to a poster campaign in local factories stressing the advantages of 'dental fitness'.[16] An increase in registration fees helped raise £1,055 from patient payments, while subscriptions totalled £900.

Despite the governors' best efforts, the following years witnessed a

[12] BCLLS, Ear Hospital, Annual Report, 1938.

[13] UBSC, Dental Hospital, Annual Report, 1920.

[14] Welshman, 'Dental health as a neglected issue in medical history', 313; Smith and Cottell, *History of the Royal Dental Hospital*, 85.

[15] *Birmingham Gazette*, 13 December 1913.

[16] UBSC, Dental Hospital, House Committee Minutes, 1917–1935, MSS 6.

decline in student numbers, income, as well as dental fitness. With student numbers dropping from sixty-five to forty-six in 1924, conservation work fell off, but extractions and prosthetics work increased. Dependent on student numbers, and especially their labour, the hospital saw its deficit increase tenfold to £1,500. With the cessation of school dental clinics in September 1923, so, too, was the Corporation's support discontinued. Possessing few investments, governors appealed for £5,000, which they invested in War Loan Stock. Despite these and other prudent measures, the debt reached £5,361 by 1925, given some overdue alterations. Besides new hot-water and ventilation systems, the hospital opened a bacteriology department under Dr Cyril Macguire.[17] While the endowment of a dental chair in 1926 helped increase investments, as well as conservation work, the debt hovered around £5,000 until 1930, when another four chairs were endowed by Elizabeth Sheffield in memory of her brother, Charles Croft.[18]

Fortunes recovered, but not as a result of a single act of benevolence. With seats on the Hospitals Council, the hospital's administration was brought into unison with the city's other medical institutions, and an additional annual contribution of £3,000 from the contributory scheme soon reduced the charity's debt to a manageable sum.[19] Not surprisingly, alterations resumed, commencing with a new roof. Besides equipping the mechanical department with an additional plaster room and augmenting the conservation room with ten chairs, the bacteriology lab, rechristened the pathology department, was moved from the first floor to the basement to make room for local anaesthetic and radiology work. While reinvestment may account for a rise in student numbers in 1932, investment invariably improved teaching facilities, if one should judge from examination results alone. Of those students sitting the LDS exams, not one failed, placing the school 'in the forefront of the teaching schools of the country'.[20] Nevertheless, fewer young persons entered the profession in these years and the work of the hospital subsequently declined, despite the appointment of a second stipendiary anaesthetist in 1936. According to the medical staff, numbers would not improve as long as local families encouraged their sons to enter trades rather than professions.[21]

The appointment of the first Professor of Dental Surgery, Humphrey F. Humphreys, in 1935, appeared to open a new chapter in the hospital's

---

[17] Ibid.

[18] UBSC, Dental Hospital, Annual Report, 1930.

[19] UBSC, Dental Hospital, Annual Report, 1932.

[20] UBSC, Dental Hospital, Annual Report, 1931.

[21] UBSC, Dental Hospital, Annual Report, 1936.

history.[22] However, its future was still uncertain. As a result, governors aligned their charity more closely with the city's other hospitals. Unlike Birmingham's other specialist hospitals, the dental charity remained the sole institution without inpatient facilities. While this provoked little anxiety in previous years, by 1937 it was accepted that the dental hospital of the future required such accommodation. As a result, when plans to build a new Hospital Centre in Edgbaston developed in the 1930s, it was decided to transfer the dental charity to the new site at a cost of £25,000.[23] Instruction in dental mechanics would be offered at the dental school, while a casualty department at Whittall Street would be worked in association with the General Hospital. Besides a £2,000 grant from the Dental Board in London and £3,000 from the University of Birmingham, the move was to be financed by the sale of the Dental Hospital, which was expected to raise £15,000. In the end, members of staff were left disappointed and would continue to occupy their former premises when plans for service integration were brought to a standstill by war. With many more medical practitioners subsequently questioning the practice of training dentists alongside medical students, the issue was never raised again.[24]

FOCUSING ON CHILDREN: THE EYE HOSPITAL

Given a healthy industrial economy, the Eye Hospital remained one of the busiest specialist institutions throughout the interwar period. However, this was least obvious in terms of inpatients, whose numbers rarely surpassed 1,800 in the 1920s and reached 2,200 in 1939. Far greater increases were evident among outpatients, who totalled 32,400 in 1920 and nearly 44,000 by 1929; similar figures were reported by Manchester's Eye Hospital in these years.[25] Increases continued the next decade, with outpatient numbers approaching 53,000, half of which were casualties. Though the cessation of munitions manufacturing initiated a decline in accident cases, the institution annually removed foreign bodies from the eyes of 18,000 locals.[26] Ever mirroring changes in local industry, the hospital's work in the 1930s increasingly involved treating chemical injuries.[27]

[22] R. A. Cohen, 'The Birmingham Dental Hospital', *Birmingham Medical Review* 20 (1958), 335.

[23] UBSC, University of Birmingham, Faculty of Medicine, Minute Book, 1929–32.

[24] W. A. Bulleid, 'The separation of dentistry from medicine', *British Dental Journal* 62, 3 (1937), 113–21.

[25] Stancliffe, *Manchester Royal Eye Hospital*, 78.

[26] BCLLS, Eye Hospital, Annual Report, 1920.

[27] BCLLS, Eye Hospital, Annual Report, 1939.

As at other hospitals, staff operated more frequently in these years following the introduction of safer surgical techniques. While operations on inpatients averaged 1,200 annually, outpatients underwent minor operations more frequently in the 1930s. Compared with 380 operations in 1920, these surpassed 2,900 in 1939.[28] Though operations at the Ear Hospital increased less dramatically between 1922 and 1938, they still doubled from 1,146 to 2,532.[29] Among inpatients at the Eye Hospital, the most common operation was for cataract, accounting for 30 per cent of procedures. However, between 1922 and 1925 those for glaucoma and strabismus (or squint) climbed steadily and, by 1928, operations for squint (195) were the most common surgical procedure.[30] With the introduction of orthoptics in the 1930s, operations for squint progressively declined in that decade. As a result, cataract operations again outnumbered all others.

Initial alterations to the hospital building were less a result of the number of cases than the type of cases treated. For example, in 1922, staff introduced five cots and three beds for nursing mothers, due to an escalation in infantile ophthalmia, a form of neonatal conjunctivitis contracted from infections in the birth canal which reputedly accounted for 30 per cent of admissions to blind asylums.[31] There was an added threat when gonorrhoea was involved, which increased noticeably during the war. Appropriately, the link between venereal disease and ophthalmia was made the subject of the Middlemore Lecture in 1922 and, the following year, forty-nine babies passed through the new infant ophthalmia ward, twenty-one accompanied by their mothers.

A number of other alterations were commenced to correspond with the hospital's centenary in 1923. Building work, however, extended over several years, given a decision to close only a section of the hospital at a time. Alterations completed in time of the anniversary were limited to the electrification of a hydraulic lift, while funds limited further redevelopment. The main task of reconstruction involved enlarging the outpatient department by absorbing the hospital's administrative offices. Separate rooms were set aside for the use of new equipment, including slit lamps, which allowed a more detailed examination of the retina, a Gullstrand ophthalmoscope, which eliminated troublesome reflections during examinations, and red-free light, which better revealed defects in tissues and nerve fibres. Wards, originally designed for privacy, were increased in order to facilitate

---

[28] BCLLS, Eye Hospital, Annual Reports, 1920, 1939.

[29] BCLLS, Ear Hospital, Annual Reports, 1922, 1938.

[30] BCLLS, Eye Hospital, Annual Reports, 1922–5, 1928.

[31] F. W. Law, *The History and Traditions of Moorfields Eye Hospital*, vol. 2 (London, 1975), 40.

nursing and cleaning. By adding a floor to the nurses' home all nurses were brought under one roof. The wards, described as light and airy, were completed in 1927, as were separate male and female operative wards and an enlarged children's ward. The following year, a private ward was converted into an operating room for septic cases.[32]

Although plans to rebuild hospital premises ground to a halt, and the interwar period generally saw 'ophthalmology in the doldrums'.[33] In Birmingham this period was characterised by an increased expenditure on new technologies. Given the successful application of x-rays across medical disciplines, interest in other rays was particularly high. For example, in 1930, new x-ray equipment and a lamp permitting the local application of ultra-violet light, already shown to possess bactericidal properties, were purchased.[34] The following year, a steam sterilising plant was put in place and, in 1934, the first ophthalmic diathermy apparatus in the region was installed, its high frequency alternating current being used to help repair detached retina;[35] previously such cases had been treated by being left alone, or by bed rest, hot-air cradles and mercury inunctions.[36] While ophthalmologists have more recently emphasised the harmful effects of ultra-violet light, staff at the Eye Hospital established a department for ultra-violet treatment under Dr Clyde McKenzie in 1928.

Largely overlooked by medical historians, UV therapy was increasingly resorted to by medical staff at hospitals during the interwar period. Local education authorities similarly promoted this form of treatment. Although none offered artificial light treatment to schoolchildren nationally in 1924, 118 of 315 authorities in England and Wales offered the therapy in 1938.[37] Convinced that light equalled vitamin D, and eager to increase patient throughputs, medical practitioners increasingly resorted to economical treatments, such as light therapies. By 1929, at the Birmingham Eye Hospital, for example, McKenzie supervised 9,000 UV baths and 2,300 local applications in three weekly sessions.[38] Four years later, premises on Edmund Street formerly rented out were transformed into a UV clinic and dental surgery with its own waiting hall and recovery rooms. A similar department opened at the Ear Hospital in 1927, where 2,239 UV cases

---

[32] BCLLS, Eye Hospital, Annual Reports, 1923–8.

[33] Law, *History and Traditions of Moorfields*, 96.

[34] BCLLS, Eye Hospital, Annual Report, 1930.

[35] BCLLS, Eye Hospital, Annual Reports, 1931, 1934.

[36] Law, *History and Traditions of Moorfields*, 85.

[37] Harris, *Health of the Schoolchild*, 76, 109.

[38] BCLLS, Eye Hospital, Annual Report, 1929.

were treated in 1938.[39] The Dental Hospital added UV apparatus to its radiographic department in 1925, as did the Children's Hospital a year earlier.[40] Treatments at the latter institution increased very rapidly from 2,300 in 1925 to 13,000 in 1928 before declining by two-thirds a decade later in the face of increasing knowledge of the part played by diet in rickets and other diseases.[41] Though used less frequently at the Children's, UV treatment remained a key therapy at the Skin Hospital, with more than 52,000 applications reported in 1938.[42] More recently, ultra-violet light has made a return to ophthalmology in the form of short wavelength laser radiation, which has enabled non-invasive eye surgery.

As revolutionary to the work of the Eye Hospital was the foundation of a squint department where patients were treated by fusion training, or remedial eye exercises, as opposed to surgery. Though a shortage of trained assistants initially restricted the work of the department, which cost between £400 and £500 to run, by 1933 staff treated 164 patients annually.[43] Thereafter, the orthoptic department expanded under Miss Evelyn Vincent into an essential hospital activity. Meaning 'straight eyes', orthoptics was developed by Claud Worth at Moorfields Hospital in London. Using an amblyoscope, which resembled a set of mounted binoculars, Worth treated 2,337 cases of squint at Moorfields between 1893 and 1903.[44] In order to maintain the interest of his young patients, his exercises employed colourful images of animals, birds and fairy stories. An interest in his findings encouraged similar experiments, all of which were suspended by war in 1914. By 1925, work in the field resumed and was led by the ophthalmologist Ernest Maddox, whose research threw much light on the function of ocular muscles. Four years later, Maddox's daughter, Mary, opened the first Orthoptics Training School at the Royal Westminster Ophthalmic Hospital. An associated professional body, the British Orthoptic Council of Ophthalmologists, was created in 1934 and held its first examinations at the Royal Westminster that summer.[45] As a result, unlike at orthopaedic hospitals, young patients at ophthalmic institutions were less likely

[39] BCLLS, Ear Hospital, Annual Report, 1938.

[40] BDHCL, Dental Hospital, Annual Report, 1925.

[41] BCLA, Children's Hospital, Annual Report, 1928, HC/BCH/1/14/11; Waterhouse, *Children in Hospital*, 107.

[42] BCLLS, Skin Hospital, Annual Report, 1938.

[43] BCLLS, Eye Hospital, Annual Report, 1933.

[44] C. Worth, *Squint, its Causes, Pathology and Treatment* (London, 1903); *Lancet*, 18 July 1903.

[45] *British Journal of Ophthalmology*, July 1934, 429.

to endure surgical cures in an age of medicine otherwise characterised by surgical progress.

Two years later, the department at Birmingham expanded yet again, taking over a room intended for the honorary medical staff. Under a new Medical Director of Orthoptics, Mr R. D. Weeden Butler, the department recorded 11,000 attendances and trained four of its own students, the majority of whom were women, for the Orthoptic Council's diploma.[46] Undertaking practical work in orthoptics, students attended lectures and demonstrations in anatomy, physiology and optics. Under the direction of Miss J. B. Lemon and an assistant, Miss Sylvanus-Jones, the department recorded nearly three times as many attendances as the orthoptics clinic at Moorfields in these years.[47] It also made many home visits and hired out its instruments for private use. Never the preserve of ophthalmologists alone, such equipment was simultaneously incorporated into the practices of opticians and would become a familiar sight on British high streets.

While staff eagerly acquired the latest medical technologies, plans for a nurses' home failed to advance, despite governors possessing a building fund that stood at £46,484 in 1935. By 1937, a plot of land at Ludgate Hill and Lionel Street was finally obtained for the purposes of a residence for sixty to seventy nurses, and the old home was converted into an inpatient department. By 1938, the original plan for a nurses' home was abandoned and rented accommodation in Handsworth acquired instead. Revealed in 1939, the home's revised plan comprised a four-story building, with accommodation for fifty-nine nurses and twenty domestic staff, as well as a lecture room.[48]

The work of the hospital pathologist also underwent great change during the interwar period. In the 1920s, when perhaps 400 tests were conducted annually, the pathologist undertook the work of anaesthetist and registrar. By 1929, when tests numbered 2,197 annually, the pathologist spent the majority of his time conducting diagnostic tests. These surpassed 3,000 in 1933 and reached 4,214 in 1939,[49] a quarter of tests being carried out in efforts to diagnose cases of infantile ophthalmia. Another quarter were conducted on patients prior to operation in attempts to separate septic cases from non-infectious ones, such as glaucoma and cataract. As a result, pathologists at the hospital and elsewhere spent far more time than in past decades diagnosing disease.

[46] BCLLS, Eye Hospital, Annual Report, 1936.
[47] Law, *History and Traditions of Moorfields*, 98.
[48] BCLLS, Eye Hospital, Annual Report, 1939.
[49] BCLLS, Eye Hospital, Annual Reports, 1933, 1939.

By 1937, the Eye Hospital also finally obtained its first convalescent home. In that year, the Colmore Estate renewed the hospital's lease to 1996, and Mr and Mrs F. W. Rushbrooke donated Burcot Grange, a nine-acre plot of land at Blackwell, for the purposes of an annex for twelve female patients and twenty children. Though no male patients were admitted to Burcot, they were sent to Blackwell, accommodated at 'the Uplands', a Hospital Saturday home. All patients travelled to Blackwell by ambulance, but were required to bring their own boots for country roads, as well as towels, flannel, soap, a brush and comb, toothbrush and slippers. During 1937, 323 attended Burcot for what must have been a relatively active convalescence.[50]

SPECIALISATION WITHIN A SPECIALTY: THE CHILDREN'S HOSPITAL

By 1920, the debt at the Children's Hospital was greater than ever before, reaching £5,451. Despite this notable shortfall at a time when subscriptions totalled £3,500, hospital administrators made every effort to exploit alternative funding opportunities and reduce expenses. For example, in order to cut costs, the governors, as in the past, closed two wards. In contrast to former closures, they now opened negotiations with administrators of the local authority's Infant Welfare Scheme in 1920 in an effort to utilise excess space. A six-cot, paying ward was also opened, increasing the institution's capacity to approximately 100 beds. Unusually, subscriptions again began to grow, bringing in almost three times the amount raised in 1916. Shortly thereafter, governors organised a conference of superintendents and teachers to encourage donations from Sunday schools, while the hospital's ten-year-old After Care Committee continued to check charitable abuse by carrying out 140 home visits in 1920.[51]

Many other initiatives developed and further demonstrate the diversification of hospital income sources in the years following the First World War. For example, in 1921, a stamp scheme was started, whereby householders collected a different penny stamp each week. Local school children also continued to compete for the Football Challenge Shield, which was donated to the hospital by Aston Villa football club, entrance fees to all matches going to the hospital. While this was predicted to raise at least £50 annually, earnings reached £175 in 1922. The hospital Brick League also continued to gain dozens of new members annually, membership rising from 136 in 1913 to 1,141 two decades later. Jumble sales were also held (1923) and became an annual event that attracted numerous mothers, who

---

[50] BCLLS, Eye Hospital, Annual Report, 1937.
[51] BCLA, Children's Hospital, Annual Report, 1920, HC/BCH/1/14/9.

were said to 'fight over the children's clothes'.[52] Patients' fees, however, raised the greatest amounts, surpassing £4,000 in 1927, then £6,000 in 1931, before halving in 1936. Only with the introduction of the Contributory Fund in January 1928 was the hospital freed from the 'nightmare of bank overdrafts'.[53] Though covering less than half the cost of patients belonging to the scheme in 1930, when contributions were raised from 2d to 3d in January 1932, the fund paid 70 per cent of its members' hospital costs despite the depression. It also reduced the hospital's deficit to a manageable £672 in 1936.

A healthy budget was especially important to governors of the children's charity at this time as its staff appeared to establish every medical specialty already represented at existing voluntary hospitals at their institution. For example, already possessing both eye and dental departments, it gained an ear and throat department (1922), as well as a venereal clinic (1925).[54] In 1927, the same year a part-time dental mechanic joined the dental department, an orthopaedic surgeon was also appointed and, in 1929, a dermatologist joined the medical staff. By 1931, the eye department opened its own orthoptic department. In subsequent years, a department of child psychology (1936) and a child guidance clinic (1937) were established and assistants were appointed to existing specialist departments.

Other changes in this period included the long-overdue reconstruction of the hospital's outpatient department in 1925. The old department at Steelhouse Lane was converted into a new massage department, and a rheumatic heart disease and chorea ward was also set up. In 1926, another ward was transformed into a twenty-seven-bed isolation ward. The following year, two more wards were considered given the demand, as was an isolation block for children with diphtheria, though this did not end outbreaks of the infectious diseases that had periodically interrupted the work of the hospital since it opened. One great advance was the introduction of routine 'Dick' and 'Schick' tests in 1934 in order to determine the susceptibility of new patients to scarlet fever and diphtheria respectively; of 5,085 patients examined in 1935, 13 per cent tested positive for diphtheria.[55] Finally, in 1928, governors announced the construction of a separate building for 100 of the hospital's most susceptible children, namely those under the age of two. While the first section with its sixty-two beds was finished in 1929, the new babies block was not opened until 1939, with little fanfare

[52] BCLA, Children's Hospital, Annual Report, 1923, HC/BCH/1/14/9.

[53] BCLA, Children's Hospital, Annual Report, 1934, HC/BCH/1/14/12.

[54] BCLA, Medical Committee Minute Book, 1921–36, HC/BCH/11/4/6.

[55] BCLA, Medical Committee Minute Book, 1935, HC/BCH/1/14/12.

owing to the outbreak of war. It was not completed until 1941 and was officially named the Leonard Parsons Block in 1947.

With this increase in beds and services, so, too, did patient numbers keep
pace with each increase in income. In 1920, the hospital was treating 1,959
inpatients and 11,585 outpatients. While this was not much more than were
seen at the Eye Hospital, its death rate remained unusually high at 13.5
per cent.[56] By 1930, inpatients had risen to 3,000, though another 3,600
attended the day wards where they recovered from adenoid operations.
Outpatients numbered 16,552, comprising 5,465 medical, 9,727 surgical and
1,360 accident cases. The hospital's death rate declined dramatically to 3.5
per cent, a feat helped by the vast number of non-fatal cases treated at
the hospital's specialty clinics. For example, in 1930, nearly 10,000 children
were attending the hospital for UV treatment, an equal number received
massage therapy, while 3,700 benefited from Swedish remedial exercises.[57]
As this should suggest, death rates are not always an adequate way to judge
the efficacy of local hospitals in these years.

ADVERTISING THE CAUSE: THE WOMEN'S HOSPITAL

By 1920, the income of the Women's Hospital reached £14,674, compared
with approximately £7,200 a decade earlier. That same year, however, it was
reporting a deficit of £4,843, and its total debt had reached £10,000. By
this time, the hospital had become far more reliant on patients' fees, which
totalled £5,030, compared with subscriptions which comprised a mere
£2,043 (14 per cent), less than donations and only slightly more than the
Hospital Saturday Fund provided the institution. Furthermore, public sector payments had risen to more than £1,700 due to work carried out by the
hospital's venereal clinic. Although fees and invested income, for example, would climb during the next decade, the board continued to devise
methods to generate public sympathy, 'because if they could get to people's
hearts they could also get to their pockets'.[58] Located outside the city centre, and away from public view since 1878, the governors, like the founder
of the Dental Hospital more than half a century earlier, believed the solution to their financial difficulties lay in better advertising the work of the
institution.

Work at the hospital in 1920 involved treating 1,800 inpatients and
performing 1,600 operations, in addition to adding some 3,290 new outpatients to a list of 16,880 active cases. These leapt to 2,750 inpatients, 3,471

---

[56] BCLA, Medical Committee Minute Book, 1920, HC/BCH/1/14/9.

[57] BCLA, Medical Committee Minute Book, 1930, HC/BCH/1/14/11.

[58] BCLA, Women's Hospital, Annual Report, 1921, HC/WH/1/10/7.

new outpatients and 18,548 total cases the following year and remained at similar levels for the next decade. The hospital was also performing approximately 1,000 abdominal sections annually, though, unlike in previous decades, the hospital's death rate remained unchanged at 2.5 per cent. In addition, another 900 patients were admitted to the Maternity Hospital each year, while 1,500 district midwifery cases were attended in their homes.[59] With each year, the work of the hospital only emphasised the risky nature of childbirth and ever greater numbers of women would deliver in the presence of a trained medical practitioner, or even in the hospital.

Alterations at this time primarily involved finishing projects that had been interrupted by war. For example, in 1921, the pathology laboratory at Sparkhill was finally completed thanks to a grant from the Feeney Trust; all pathological and bacteriological work could now be carried out in the hospital. In 1921, this involved conducting 410 tests, the majority (249) on tumours, and this steadily rose to 2,400 examinations by 1930, 1,479 on tumours alone.[60] In the immediate post-war years, much additional work involved treating approximately 150 new venereal patients at an evening clinic that opened in July at the request of the Ministry of Health (and closed in June 1925). Alongside the pathology lab at Sparkhill, staff also opened a new chapel and mortuary, where the bodies of forty-two women were examined in the year of its completion (1922). Besides the addition of sixteen beds, nurses' quarters, lecture and meeting rooms in 1924, all other changes involved improving the facilities of the hospital's associated charity, the Maternity Hospital.

An extension of the Maternity Hospital at Loveday Street, estimated to cost £22,000, was announced in 1922 and promised greatly to increase the charity's status.[61] An agreement with the General Hospital to transfer its maternity department to Loveday Street ensured that many more students would receive midwifery training at the charity in efforts to reduce a national maternal mortality rate that remained obstinately high at around 4 per thousand births in the early 1920s.[62] As predicted, while only six medical students attended the hospital in 1913, fifty-eight were recorded in 1925, the year the extension was completed.[63] Maternal mortality, on the other hand, increased by 20 per cent over the following decade. Not surprisingly, the work of the Maternity Hospital at this time began to be regarded as more important than that of the Women's Hospital, as it not only trained

---

[59] BCLA, Women's Hospital, Annual Report, 1923.

[60] BCLLS, Women's Hospital, Annual Reports, 1921, 1930.

[61] BCLA, Women's Hospital, Annual Report, 1922, HC/WH/1/10/7.

[62] Hardy, *Health and Medicine in Britain*, 95.

[63] BCLA, Women's Hospital, Annual Report, 1925, HC/WH/1/10/7.

students, midwives and mothers, but it also reduced the work of the parent institution. With sixty-three beds, the hospital had doubled in size within a decade and ante-natal clinics were held each morning instead of twice a week. A new operating theatre and pathology lab were also planned, while funding from the Feeney Trust finally permitted the hospital to acquire its own x-ray apparatus in 1927.[64]

By 1930, beds at both hospitals numbered 182, up from 130 a decade earlier. With only a quarter of their £37,000 combined income coming from the Hospital Contributory Scheme, the number of paying beds was increased. More reliable, patients' payments comprised a third of income, while subscriptions comprised only 8 per cent of income, compared with 14 per cent a decade earlier. Local authority funding at the Women's had declined to £1,300, but the Maternity Hospital was also now in receipt of several special grants from the public sector. Of 4,300 inpatients, 60 per cent, nearly all operative cases, attended the Women's Hospital. The majority (1,516, or 87 per cent) turning up at the Maternity Hospital were abnormal deliveries, nearly a quarter of these were emergencies sent by outside doctors. Outpatients numbered 6,877; 4,646 attended the Women's Hospital on nearly 17,000 occasions. Of the Maternity's 2,231 outpatients, 1,643 were delivered by midwives and medical students in seven local districts. In total, patients made nearly 14,000 attendances, and 488 additional women attended a baby clinic on 1,670 occasions.[65] As a result of such work, the interwar years saw a remarkable shift in the location of childbirth from the home to the hospital, with a quarter of all births in 1940 taking place in medical institutions.[66] A decade later, the percentage of total births confined in the region's hospitals alone rose to nearly 56 per cent.[67] Significantly, women's chances of surviving childbirth improved only with the development of new drug therapies and not increases in medical intervention.

During successive years, patient numbers and expenditure climbed, though in 1932 governors reported an improved financial position. Thanks largely to nearly £13,000 from the local contributory scheme that year, against £9,000 in 1931, the hospital's deficit declined from £5,000 to £3,000.[68] Faced with lower debt, a new nurses' home for the Maternity Hospital was commenced and finished the following year. In 1934, a new

[64] BCLA, Women's Hospital, Annual Report, 1927.

[65] BCLA, Women's Hospital, Annual Report, 1930, HC/WH/1/10/8.

[66] Hardy, *Health and Medicine in Britain*, 105.

[67] [Birmingham Regional Hospital Board], *Birmingham Regional Hospital Board*, 67.

[68] BCLA, Women's Hospital, Annual Report, 1932, HC/WH/1/10/8.

radium ward opened in Sparkhill, a surgeon radiologist, Miss Wilmott, managing the radium, which the hospital purchased for £4,500;[69] 176 patients, most with ovarian cancer, underwent radium treatment the following year. In 1936, a maternity 'flying squad' was established to tackle maternal mortality in the city. Nationally, maternal mortality witnessed its greatest fall in these years, declining from 4.2 per thousand births in 1936 to 2.2 per thousand in 1940.[70] Working in association with the Birmingham Health Committee and the Contributory Association Ambulance Service, the squad, which assisted two dozen cases annually, was pioneered by Hilda Lloyd, who joined the staff of the Women's Hospital in 1921 and ran its post-war venereal clinic.[71] Appointed lecturer in midwifery and diseases of women at the University of Birmingham in 1934 and rising to professor in 1944, Lloyd was elected the first female president of the Royal College of Obstetricians and Gynaecologists in 1949.[72]

In 1938, despite the work of the flying squad, the region served by the institutions contracted. As the result of the Midwives Bill in 1937, governors decided to concentrate on the five districts closest to the hospitals; two additional districts formerly served by the women's hospitals were assisted by the city's Public Health Committee. Inpatient numbers increased to 5,232 and were distributed almost equally between the institutions. Nearly two-thirds, or 4,269, of outpatients, however, attended the Women's Hospital. Given the popularity of its ante- and post-natal clinics on the other hand, more than half of all attendances (24,284) were at the Maternity Hospital. With expenditure under control, governors commenced an appeal for £125,000 to build a new 125-bed Maternity Hospital on the Loveday Street site.[73] Unfortunately, as occurred at many hospitals a generation earlier, only half of the estimated building costs were raised by the time a national crisis once again interrupted collections.

A JOINT EFFORT:
THE ORTHOPAEDIC HOSPITAL

With twenty special military orthopaedic centres established in Britain by 1918, orthopaedic surgery was given greater prominence during the Great War, and both developments in surgery and demand from injured veterans

[69] BCLA, Women's Hospital, Annual Report, 1934, HC/WH/1/10/9.

[70] Hardy, *Health and Medicine in Britain*, 105.

[71] BCLA, Women's Hospital, Medical Committee Minute Book, 1893–1928, WH/1/5/2.

[72] V. E. Chancellor, 'Lloyd, Dame Hilda Nora', in *Oxford Dictionary of National Biography*.

[73] BCLA, Women's Hospital, Annual Report, 1938, HC/WH/1/10/9.

underlined the importance of rebuilding existing orthopaedic hospitals.[74] As two-thirds of all war casualties suffered locomotive injuries, bed numbers failed to keep up with demand.[75] The work of the Birmingham charity similarly continued to be hampered by inadequate accommodation in the immediate post-war years. The long periods required to treat some conditions ensured extensive waiting lists, and expenditure continued to increase more rapidly than income, a situation that had become common at all institutions, regardless of specialty. As a result, cost-cutting measures were introduced, including the appointment of an almoner (1921). While a Linen League continued to dress all hospital beds, a workshop staffed by former patients began to manufacture and repair surgical appliances in 1921.[76] Despite such efforts, the hospital's income fell £3,000 short of expenditure the following year, largely due to a decline in the local economy. In future, greater state spending in the form of education grants and other payments would make up such deficits. Outbreaks of infectious disease, on the other hand, closed several wards and further reduced the work of the charity.

Demands on the hospital increased and necessitated additional staff. In 1923, the post of honorary assistant surgeon was revived and a physician was also appointed to cope with the number of patients suffering from diseases of the nervous system. Additional costs encouraged more efficient co-ordination with existing medical services, commencing with a decision to close the 1923 financial year on 31 December, in order to fall in line with the city's principal hospitals. A joint School of Massage was then formed with the General Hospital. Visiting the homes of individuals on the hospital's waiting list, the almoner negotiated assistance from outside agencies to enable surgeons' instructions to be carried out. Under a new scheme, the hospital received 15 shillings from the City of Birmingham for each case of surgical tuberculosis treated and greater use of the beds at Moseley Hall Convalescent Home eased the pressure of treating long-term cases. Though plans for the charity's own convalescent home were shelved until finances improved, the generosity of a hospital committee member placed a house at Vicarage Road at the disposal of the institution. Opened in 1925, its thirty beds freed as many places in the main Newhall Street building.[77]

A more significant improvement in the charity's circumstances was the

74 R. Cooter, *Surgery and Society in Peace and War: Orthopaedics and the Organisation of Modern Medicine, 1880–1948* (Basingstoke, 1993), 105.

75 Borsay, *Disability and Social Policy*, 51–2.

76 BCLLS, Orthopaedic Hospital, Annual Report, 1921.

77 BCLLS, Orthopaedic Hospital, Annual Report, 1925.

result of co-operative effort. In 1924, governors announced their intention of amalgamating with the Birmingham Cripples' Union.[78] Begun in 1896 as a department of the Hurst Street Mission, the Cripples' Union became a distinct charitable organisation in 1899, assisting 330 disabled children on an annual income of £365.[79] Two decades later, income approached £30,000 and it assisted 1,500 children and 500 adults. Run by a general secretary, Frank Mathews, the Cripples' Union offered preventative and curative care, as well as vocational training. Possessing its own convalescent home since 1900, the Union and its associate branches in Smethwick (1903) and Selly Oak (1905) sent many more children to independent homes across the country. The charity's curative work improved significantly in 1909 when the Woodlands on the Bristol Road, donated by the late George Cadbury, opened its thirty-seven beds to disabled children. The same year, the children's charity merged with the Adult Cripples' Guild and the organisation was renamed the Birmingham and District Cripples' Union.

Prior to amalgamation, the Cripples' Union had provided some of the Orthopaedic Hospital's patients with spinal carriages and even admitted cases of surgical tuberculosis to the Woodlands. With amalgamation on 31 March 1925, however, the enlarged orthopaedic charity with its 257 beds, was spread over four sites. In addition to the Orthopaedic Hospital's fifty-bed Newhall site and thirty-bed convalescent home in Edgbaston, the Cripples' Union brought twice as many beds to the scheme, with ninety-seven beds at the Woodlands in Northfield and another eighty at the Forelands open-air convalescent home in Bromsgrove (which had been acquired and equipped by the charity in 1920 for £30,000, half of the funds coming from the Ministry of Health). The merger transformed the charity into the third largest hospital organisation in the city.[80]

Soon the amalgamated charities acquired a fifth site, but a process of consolidation had also started. Using funds left by John Avins to the Orthopaedic Hospital, the charity purchased the old Children's Hospital in Broad Street in 1925. As the charity's outpatient department, it held all electrical work, a massage department – staffed by a head masseuse and sixteen assistants and under the control of Dr S. A. MacPhee – and newly acquired artificial sunlight treatment were transferred to the site. Bootmaking workshops also moved there from Islington Row and Lionel Street. By 1927, outpatients no longer attended Islington Row and Newhall

---

[78] BCLLS, Orthopaedic Hospital, Annual Report, 1924.

[79] [BCU and ROSH], *Birmingham Cripples Union and Royal Orthopaedic and Spinal Hospital, 1817–1915* (Birmingham, 1925), 8.

[80] Ibid., 3.

Street. With a third of hospital patients residing outside Birmingham, external clinics were established at Nuneaton, Dudley, Stourbridge and Walsall.[81] Initially, hospital staff attended these clinics once a month, but many appointed their own officers and eventually attained independent status.

With outpatient work consolidated at Broad Street, inpatient treatment was concentrated at the Woodlands in Northfield. The extension, which cost more than £24,000, added 100 beds to the institution (sixty for children and forty for adults), as well as a nurses' home and recreation hall in order to entertain a nursing staff of sixty-five women who resided 5 miles from the city centre. Given the greater presence of adult patients, male orderlies were appointed to help with lifting, though they were subsequently discovered to exert a good 'disciplinary influence' on the wards. With the closure of the Newhall Street site, the wards became even more crowded. By the time the Vicarage Road site closed in 1938, the Woodlands had 207 beds and as many patients on its waiting list. During that year, 1,461 inpatients stayed at the hospital, 1,005 operations being performed.[82] With isolation wards and a decision by the Medical Officer of Health to immunise all children against diphtheria in 1934, the work of the hospital, as at the Children's, was less often interrupted by outbreaks of infectious diseases.

While Birmingham's voluntary hospitals had long claimed to rehabilitate the city's sick and injured workers, the Orthopaedic Hospital could legitimately claim to be like no other institution when it came to producing 'productive' citizens. In 1926, its vocational courses for girls and boys over sixteen years of age were approved by the Board of Education. Based at the Forelands, courses for girls taught weaving and machine knitting, while boys were instructed in poultry work. Though appropriate to their rural location, more marketable skills, such as carpentry, were introduced in 1929 and training for girls incorporated typing and shorthand lessons.[83] Other hospitals commenced similar schemes, training children as shopkeepers, though perhaps benefited less directly from their resident labour force. Training, nevertheless, was similarly gendered.[84] At Birmingham, the carpentry class undertook all alterations to hospital buildings, while workshops employed former patients to make boots and surgical instruments. In 1930, its twenty-seven employees constructed more than 3,000 boots and splints, repaired nearly 15,000 instruments and began to supply

[81] BCLLS, Orthopaedic Hospital, Annual Report, 1927.

[82] BCLLS, Orthopaedic Hospital, Annual Report, 1938.

[83] BCLLS, Orthopaedic Hospital, Annual Report, 1929.

[84] *Lancet*, 7 December 1929.

other institutions. Another fifty discharged patients produced handicrafts, including ladies' lingerie, which won prizes at international exhibitions and were sold from premises on Broad Street, raising approximately £1,400 annually in the 1930s. From 1930, a committee began to co-ordinate the various training departments and regularly liaised with regional disability organisations.[85]

Visiting the 'hospital in a garden' in 1932, George Newman, the chief medical officer of the Ministry of Health, regarded the charity as the 'most perfect example to be found in this country or elsewhere of voluntary enterprise'.[86] Further extension schemes were approved in 1934 for private wards and another appeal was launched in 1937 to facilitate the transfer of patients to the Woodlands from Vicarage Road, which was closed in 1938.[87] Though the latter appeal proved disappointing, raising only £28,000 in its first year, the charity was never again crippled by debt. Not only did patients help raise funds in a variety of ways, but all patients, unless one of the 10 per cent who were sponsored by the public authority, paid a shilling registration fee and private wards were filled to capacity throughout the next year. Demand for orthopaedic services had by 1936 led to the establishment of another thirty-nine orthopaedic hospitals nationally, as well as another 400 specialist clinics.[88] So influential to the specialty's further development, war disrupted these institutions' and the Birmingham charity's work by reducing bed numbers, delaying further construction and dispersing patients still further into the countryside and away from the dangers of enemy action.

INDUSTRY AND ECZEMA:
THE SKIN HOSPITAL

Having secured land on which they intended to construct a new hospital, the governors of the Skin Hospital outlined plans for a fifty-bed institution in their 1920 annual report. Despite their vision and the acquisition of a plot of land for £7,000, plans failed to advance. Critical of the city's elite, who were said to be lacking 'the right sense of values',[89] the governors concentrated instead on refurbishing their Bright Street building.

Treating 381 inpatients and 7,289 outpatients, who made 67,225 attendances in 1924, the hospital's therapies had altered since the first decades of the century. In 1921, an 'Erlangen [radiographic] apparatus' was acquired

[85] BCLLS, Orthopaedic Hospital, Annual Report, 1930.

[86] BCLLS, Orthopaedic Hospital, Annual Report, 1932.

[87] BCLLS, Orthopaedic Hospital, Annual Reports, 1934; 1937–8.

[88] Borsay, *Disability and Social Policy*, 53.

[89] UBSC, Skin Hospital, Annual Report, 1924.

to treat deep-seated cancers.[90] Used on 124 cases two years later, the treatment was praised for its mathematical accuracy, and was compared with artillery soon after its acquisition; in subsequent years staff would often employ military metaphors to describe their fight against cancer.[91] While an electrical department treated nearly 11,000 cases annually, and commenced UV treatment in 1925, 645 patients still made use of the hospital's less technologically advanced bathing facilities.[92]

Despite a rise in venereal cases during the war, by 1925, 94 per cent of patients were described as skin cases, and evening venereal clinics were discontinued.[93] State-aided treatment centres were said to have reduced the incidence of venereal disease, and dermatitis, unheard of in previous years, now afflicted thousands of local workmen, especially in nickel and chromium-plating trades.[94] By 1930, these changes were reflected in a decision to change the name of the hospital to the 'Birmingham and Midland Skin Hospital'. Greater interest in the health of occupations led hospital staff to co-operate with the newly created Department of Industrial Hygiene at the University of Birmingham. By 1937, annual reports, like those of the Eye Hospital, more often emphasised the hazards of chemicals processes.

Unlike the Eye institution, and still tainted through its association with venereal disease, the Skin Hospital struggled to raise its income. With a debt of £16,000 in 1926, governors gave up plans for a new hospital and, like the dental institution, sought to relocate to the new Hospital Centre at Bournbrook.[95] After the local contributory scheme helped governors reduce the hospital's deficit to £1,000 in 1929, nothing further was heard of this relocation plan. Other innovative fundraising tactics included placing full-page advertisements in annual reports and approaching the mayors of the numerous country districts that the charity continued to serve.[96] By 1930, governors again discussed rebuilding in the country. An extension scheme was commenced, but reluctantly suspended the following year.

Their appeal was recommenced in July 1933. Having sold its rental properties on Beak Street after they were condemned by the City Council, the hospital started an Old Patients Association and a tea canteen for

---

[90] The hospital's archives reveal little else about this technology, which was presumably developed by Siemens at their factory in Erlangen, Germany.

[91] See, for example, S. Sontag, *Illness as Metaphor and AIDS and its Metaphors* (London, 1990), 64, which considers the military metaphors associated with cancer treatment.

[92] UBSC, Skin Hospital, Annual Report, 1925.

[93] Ibid.

[94] UBSC, Skin Hospital, Annual Report, 1930.

[95] UBSC, Skin Hospital, Annual Report, 1926.

[96] UBSC, Skin Hospital, Annual Report, 1932.

outpatients. The same year, work commenced on an eighteen-room nurses' home at 35 George Street in Edgbaston, the former nursing home of Christopher Martin, surgeon to the Women's Hospital, who died in 1932. Alongside the home, governors constructed a three-story, fifty-bed block with sixteen beds and three private wards for children. Opened in July 1935 by the Earl of Dudley, the building cost £32,129, outstanding debts totalling £4,770.[97] In 1936, with half of the hospital's £14,000 income coming from the local contributory scheme, governors spent another £3,876 on improving their outpatient department at John Bright Street.[98] By 1938, with the concentration of deep x-ray treatment at the General Hospital, the hospital ceased such work and concentrated on skin therapies and more traditional forms of treatment.

A ROYAL DEFICIT:
THE QUEEN'S HOSPITAL

With a deficit nearly the size of its £24,000 annual income, the Queen's Hospital was in as precarious a position as the city's other voluntary hospitals in 1920. Occupying premises long in need of alterations ensured its debt would only further accumulate. Rising costs in these years alone were due to the installation of new heating apparatus, post-war cleaning of the building and a bacteriological department, where Mr R. B. H. Wyatt conducted 620 tests, including 107 Wassermann examinations for syphilis in his first year as bacteriologist.[99] Other additions included Birmingham's first electrocardiograph. Used sixty-four times in its first year, and another 200 times in 1921, the machine was donated as a memorial to members of staff who died during the war.[100] Its medical outpatient department and operating theatres, however, were unchanged since 1873 and 1841 respectively, and their refurbishment would have to wait until finances improved.

The following year, the deficit was reportedly halved, due to new forms of income, lower costs and improved methods of purchase, which reduced the average cost per bed nearly 20 per cent in a single year to £161.[101] New income included a one-off grant of half a million pounds which the government, as advised by the Cave Committee (1921), distributed between the nation's voluntary hospitals in order to meet their war deficits.[102] More lasting relief came with a decision to charge patients coming from outside

---

[97] UBSC, Skin Hospital, *Annual Report*, 1935.

[98] UBSC, Skin Hospital, *Annual Report*, 1936.

[99] BCLLS, Queen's Hospital, *Annual Report*, 1920.

[100] BCLLS, Queen's Hospital, *Annual Report*, 1920–1.

[101] BCLLS, Queen's Hospital, *Annual Report*, 1921.

[102] Abel-Smith, *Hospitals*, 298.

the Birmingham and Smethwick area £3 3s a week for treatment, £800 being raised in the first year of the scheme. A similar amount was raised by a carnival, which was organised by local medical students and became an annual event. In addition, a hospital clothing fund as existed at other hospitals was commenced, supplying more than 300 garments in 1922.

Though patient numbers declined during a brief period of local unemployment, by 1923, the hospital's surgical block had become 'entirely unsuitable for the surgical work of a modern hospital'.[103] Unlike in past decades, construction only began after the Joint Hospitals Committee approved governors' plans to convert the 'cramped and uncomfortable' portion of the building. Opening their 120-bed surgical extension in 1927, governors described their institution as one of the most up-to-date hospitals in the UK.[104] As its 100 new beds required the services of forty-four nurses, a further round of alterations involved the construction of an additional nurses' home. Though successfully retaining the 'homely, personal touch of [its] staff', the hospital acquired a debilitating debt.[105] With the new surgical wards costing an extra £15,000 a year to run, the hospital accumulated a £22,000 overdraft by 1931. Reluctantly, and somewhat controversially, the Joint Hospitals Committee permitted the Queen's, together with the General and Ear hospitals, to make a joint public appeal for £90,000.

Commencing the post-war period with 3,229 inpatients and 19,128 outpatients, by 1930, the patient population rose to more than 5,000 and 32,000 respectively. While some of this increase was due to the hospital's extension, it was also the result of new clinics and specialist departments. For example, a department of psychotherapy was commenced in 1920 and plans to affiliate with the Nerve Hospital in Bath Row, which opened in 1919, became reality in 1923. Besides sharing the expenses associated with a massage department and bacteriological laboratory, the Queen's Hospital provided its associate institution with the anaesthetists required in cases of neurological surgery.[106] A new ENT department and the appointment of a dental house surgeon at the Queen's also led to a less costly increase in outpatients. As a result, while staff saw only 225 eye cases in 1928, they treated 1,126 ENT and 2,452 dental cases that year. While the former department gained five beds in 1932, the dental department gained a speech clinic that same year. Further additions included an allergy clinic, established alongside the bacteriological laboratory in 1935, and a fracture clinic which was enlarged in 1935. Treating 2,000 cases the following year, this particular

[103] BCLA, Queen's Hospital, General Committee Minutes, 1917–25, HC/QU/1/1/8.

[104] BCLA, Queen's Hospital, General Committee Minutes, 1925–31, HC/QU/1/1/9.

[105] Ibid.

[106] BCLA, Queen's Hospital, General Committee Minutes, 1917–25, HC/QU/1/1/8.

clinic would play an increasingly important role at the Queen's Hospital site in subsequent years.

CO-ORDINATING SERVICES:
THE GENERAL HOSPITAL

Staff at the General commenced the 1920s with fewer financial worries. A £17,800 grant from the National Relief Fund, added to the sale of securities (£40,014), the proceeds of Hospital Saturday (£9,000), Hospital Sunday (£,834) and subscriptions (£12,572), left the institution free of debt.[107] The establishment of a General Hospital League among local working men and women in order to rid patients of maintenance charges and a total of fifty-nine endowed beds, many commemorating individuals killed in the war, appeared to promise a better future. However, this would prove to be the hospital's only deficit-free year for the remainder of the decade.

The hospital started the interwar period with 6,930 inpatients and 42,376 outpatients, and staff conducted 5,923 operations annually.[108] By the end of the 1920s, inpatient numbers had changed little, totalling 7,057. Outpatients, however, now numbered 57,838 and staff conducted more than 10,000 operations annually.[109] For many local inhabitants, hospital treatment no longer implied entering an institution. In this sense, their relationship with these institutions differed very little from that which formerly existed between patients and dispensaries in the nineteenth century.

Due to the work of the hospital almoner, however, far more was known about these patients than in previous decades, especially in terms of who could pay. In fact, governors argued that the only details that doctors learned about patients' lives, home conditions and work was due to the work of the almoner.[110] Besides interviewing 1,363 new patients in 1920 to determine whether they were deserving of charity, the almoner also decided which cases required extra help.[111] For example, this same year, 381 individuals were sent to convalescent homes, thirty-five received surgical appliances, another eight obtained dentures, while 289 went to the Jaffray Hospital, the institution's convalescent home. Despite lasting concerns regarding charitable abuse, only eleven cases were found able to pay

[107] BCLA, General Hospital, Annual Report, 1920, GHB 438.

[108] Ibid.

[109] BCLA, General Hospital, Annual Report, 1929, GHB 439.

[110] BCLA, General Hospital, Annual Report, 1926, GHB 438.

[111] For more on hospital efforts to determine the 'deserving poor', see J. Reinarz, 'Investigating the "deserving" poor: charity, discipline and voluntary hospitals in nineteenth-century Birmingham', in *Reconfiguring the Recipient: Historical Perspectives on the Negotiation of Medicine, Charity and Mutual Aid*, ed. A. Borsay and P. Shapely (Aldershot, 2007).

for treatment. The post also involved co-ordinating the services of numerous other local charities in order to avoid unnecessary overlap of charitable activity. For example, of the 1,363 patients examined in 1920, forty were referred to other hospitals, dispensaries and panel doctors, another seventy-six to the Poor Law guardians. Many more were directed to the Citizen's Society (83), Cripples' Union (38), Hospital Saturday Fund (46), health department (54) and education committee (51).[112] By 1934, the contributory scheme removed the necessity for much of the financial assessment, though the almoner was still interviewing more than 600 cases annually. Her work included sending patients to convalescent homes (504), spas (35), nursing homes (10) and providing many others with surgical instruments (247), dentures (27), groceries (60) and return fares upon discharge from hospital (157).[113] Other interesting examples of the varied work undertaken by the almoner included the arrangement of adoptions and the allocation of foster parents. More unusually, two girls were sent to a training home in Liverpool and a boy was sent to a farm in Canada in 1927 following recovery.

The work of the nursing department also developed, much of it concentrating on raising the standards of the department's trainees. Similarly, the history of the nursing profession up to this time was largely concerned with the creation of a recognised qualification. Training, largely associated with the work of Florence Nightingale, commenced in the mid-nineteenth century, her school alone having certified nearly 2,000 nurses following a year's training between 1860 and 1903.[114] There were, however, more than 65,000 nurses in English hospitals and homes, and training varied with each institution. With the foundation of the British Nurses' Association in 1887, a register of nurses and their training became a central concern of the profession. Though midwives gained their own register in 1902, many remained opposed to state registration of nurses for nearly another decade. Nevertheless, the Society for the State Registration of Nurses (1902) and the BNA ensured the subject was debated regularly during the years leading up to 1914. Reform when it came was largely prompted by the war and the entry of many untrained women to the profession, the largest addition to regular nurses coming from Voluntary Aid Detachments.[115] As a result, regular nurses were anxious to register their skilled status in order to acquire a formal qualification, something the middle-class VADs lacked. The same year, the Nurses Registration Acts were passed by Parliament to

---

[112] BCLA, General Hospital, Annual Report, 1920, GHB 438.

[113] BCLLS, General Hospital, Annual Report, 1934.

[114] B. Abel-Smith, *A History of the Nursing Profession* (London, 1960), 24.

[115] Ibid., 85.

create a formal nursing register, which opened in November 1921. In 1925, the first state examinations were held and more than 4,000 nurses were admitted to the register.[116]

In Birmingham, a sister tutor was appointed at the General Hospital's nursing school in 1920 to take charge of probationers' studies.[117] That same year, a prize for the nurse with the highest marks in gynaecology was introduced (by Mr H. Beckwith Whitehouse). Further investments in teaching included a model heart and lantern slides which were purchased out of funds the nurses themselves raised. The board, however, paid a far greater sum in 1923 to acquire premises known as the Vicker's Home in Lench Street as an adjunct to the nurses' home. While this may have attracted more women to the hospital's nursing school, this was also attributed to the work of the newly founded General Nursing Council (1920) and the fact that other professions open to women were overstocked. A Preliminary Nurse Training School was subsequently opened in January 1925, probationers attending various elementary courses over six weeks. Approved by the General Nursing Council as a training school for nurses in 1922, the Queen's Hospital similarly trained many more of its nursing staff in these years. In the first year at the General, forty-four commenced training, twelve left during the first six weeks, one was dismissed and seven were still on probation at the end of the year.[118] Those entering the school enjoyed far greater recreational opportunities, including an annual tennis tournament and an inter-hospital swimming club. Not surprisingly, given these facilities and extracurricular activities, forty-nine candidates entered the school the following year, though eleven still abandoned training during the year. Besides tennis and swimming, nurses studied anatomy and physiology, as well as a full course on hygiene, which included trips to model farms, dairies, industrial sites and water works. While the hospital board hoped that improved facilities would attract 'the right type of girl',[119] the programme continued to attract the right number of girls, fifty-seven entering in 1927 and sixty-two the following year; the school also began to attract nurses from abroad, including Switzerland, France and Ceylon.[120] By 1930, eighty young women were entering the school annually. All teaching and examinations were held at Bethany House. Ear and eye prizes

---

[116] Ibid., 113.

[117] For the period leading up to these belated innovations, see S. Wildman, 'The development of nursing at the General Hospital, Birmingham, 1779–1919', *Nursing History* 4, 3 (1999), 20–8.

[118] BCLA, General Hospital, Annual Report, 1925, GHB 438.

[119] BCLA, General Hospital, Annual Report, 1926.

[120] BCLA, General Hospital, Annual Report, 1927–8.

further suggest teaching was becoming more specialised, while training fees were increased to 5 guineas in 1933 to cover additional equipment and facilities.

In contrast to the nursing scheme, the General's school for massage had a shorter history. In 1920, Miss D. I. West resigned as head masseuse after ten years to take up private work, though continued to instruct at the school, which raised more than £300 from fees, £420 in 1924. The school of massage, medical gymnastic and medical electricity opened on 3 April 1911, quickly attracting its first three students, two from among the hospital staff. By the outbreak of war in 1914, numbers had doubled and, since 1911, students, including many nurses, prepared for the exam of the Incorporated Society of Trained Masseuses. In 1916, an examination in medical electricity was also introduced. The following year, an assistant teacher was appointed to help instruct in medical gymnastics. When the Incorporated Society became the Chartered Society of Masseuses and Medical Gymnastics the training period extended to twelve months, and student numbers declined with the greater commitment required. By 1925, their numbers recovered and twenty-one nurses enrolled at the school, only a few of each year's intake remaining to train in medical electricity after passing their first exams. By 1928, twenty-eight enrolled and massage fees raised £651. The following year, though pupils' numbers dropped to fourteen, the surgical casualty officer now lectured to the nurses on the subject of fractures. By 1930, numbers increased, new students numbering twenty-four, forty-six women in total having completed at least one of the schools three main courses.[121] At this time, a similar school was commenced at the Queen's Hospital, which enrolled thirty-five students in 1932 and collected nearly £1,000 in student fees. By 1934, students and staff there were providing patients with 2,638 massages, compared with 734 in 1920.[122] Training at the Maternity Hospital similarly developed, but massage was practised there only to aid lactation. The success of the Queen's programme, on the other hand, perhaps influenced developments at the General, where numbers of pupils had begun to decline by 1932, when only one course was given. In 1933, numbers further reduced and by 1934 the school temporarily closed, though massage therapy would continue to transform medical care at local hospitals.[123]

Alterations at the city's leading hospital in this period began with less sophisticated, but costly, technological innovations. In 1920, staff acquired new boilers, and decided to draw water from the town's supply, as well

---

[121] BCLA, General Hospital, Annual Report, 1920–30.

[122] BCLLS, Queen's Hospital, Annual Reports, 1930–4.

[123] BCLA, General Hospital, Annual Reports, 1932–4, GHB 439.

water rapidly corroded the steel of earlier boilers. The following year, funds were once again in short supply, and the hospital's debt reached £17,366.[124] As a result, the annual cleaning of the hospital was cancelled and painting of the building was also rationed. Governors suggested patients contribute in some way to their treatment, though no more than 2 guineas each. A year on and it was argued that the hospital was becoming increasingly inadequate given both advances in science and medicine and the growth of the city. A grant of £8,000 from the Voluntary Hospitals Commission was required to reopen eighty beds at the hospital, and the Jaffray was also temporarily closed. An increase in legacies the following year, as well as donations from certain businesses allowed the purchase of an electrocardiograph, bone-grafting apparatus, diathermy apparatus, an x-ray transformer and an ultra-violet lamp. The installation of electric lighting at the Jaffray Hospital raised the convalescent home's deficit to £3,000, which was again carried by the parent institution.[125]

By 1924, the General's governors were preparing for expansion when they purchased St Mary's churchyard and gained permission to develop the church and its burial grounds. With more than 7,000 operations being performed at the hospital annually, new operating theatres were also required. Another £1,000 was also spent in obtaining equipment to deal with additional laundry. At this time of increased expenditure, the governors were also faced with a gamble, having been offered an increased grant from the Hospital Saturday Fund on condition that they abolish their ticket system. By this time, workers' contributions far surpassed those of local businessmen. The board naturally regarded an extension of the contributory scheme throughout the district as the only answer to their financial problems, the deficit having again surpassed £12,000 due to a decline in legacies. Neither were patients' payments to be relied on, charges in 1925 having raised only £1,805. Nevertheless, the hospital extension was postponed in 1926, as its overdraft had reached £25,000. Instead, a temporary ward with twenty beds was erected on the hospital grounds. With the contributory scheme shortly to come into effect, however, confidence was still high and, in 1927, bodies were being removed from the vaults below St Mary's in anticipation of resumed building. Street accidents involving automobiles were adding to the hospital's burden and 3,000 urgent cases were sent to other hospitals that year. In the meantime, a carnival was arranged by local workers and raised £600 for the hospital. Alterations were limited to a covered balcony for Ward 2 in memory of George Heaton, a former JP.[126]

[124] BCLA, General Hospital, Annual Report, 1921, GHB 438.

[125] BCLA, General Hospital, Annual Reports, 1920–2.

[126] BCLA, General Hospital, Annual Reports, 1924–7, GHB 438–9.

With the hospital's 150-year anniversary approaching, the time appeared ripe for expansion. By 1928, the development of the St Mary's site was estimated to cost £25,632. New wards, a casualty department and nurses' accommodation would cost another £40,000. To finance this, thirty paying beds were set aside in two existing wards. Fees totalled £3,670 in 1929. Anniversary celebrations that year, which included medical history lectures, religious services and lectures by the British abdominal surgeon Lord Moynihan and the Austrian professor and digitalis advocate K. F. Wenckebach further concentrated attention on the charity while redevelopment was in full progress. By 1930, with building work complete, the hospital had a new casualty department, three wards (containing sixty beds), a radium curator's lab, a radium store and office, a diet and teaching kitchen, verandahs for two further wards, a larger boiler house and a garage for the ambulance donated by Sir Herbert Austin. Nurses gained a recreation hall, thirty bedrooms, classrooms and additional sitting rooms. Although patient's fees raised £4,024, and the hospital possessed 108 dedicated beds, the overdraft stood firm at £20,000, and the Jaffray's deficit added another £4,600.[127] When the debt reached £52,582 the following year, it was decided to launch a joint appeal for £90,000 with the Queen's Hospital in January 1932; £30,000 had already been raised. Further relief was to come by increasing the number of paying beds to fifty, while raising workers' weekly contributions to the contributory scheme from 2d to 3d. An independent investigation into the administration of the hospital was made by a well-known firm of chartered accountants with the help of Mr R. H. P. Orde of the British Hospitals Association to ensure the greatest possible economy was practised, while the board resisted the closure of beds and other services. Although a decline in expenditure was noted, the hospital was rescued by some timely legacies, £25,000 alone coming from the estate of Mr Hugh Morton.[128]

As one might expect, the hospital made greater efforts to avoid duplication of services in these years of financial difficulty. This started with the appointment of an almoner, whose department functioned much like a clearing station, sending forty to one hundred patients to other hospitals annually. In 1923, this went a step further when the Maternity Hospital took over the General's external midwifery department in order to avoid further overlap in provision. In turn, staff at the General began to undertake all of the biochemical work of the Maternity Hospital in 1932. However, far more important links had begun to form between the city's two main teaching hospitals. Three years earlier, liaison officers were appointed on the medical

---

[127] BCLA, General Hospital, Annual Report, 1930, GHB 439.
[128] BCLA, General Hospital, Annual Reports, 1931–2.

committees of both the Queen's and General Hospital in order to unite policy.

The idea for a Hospitals Centre in Birmingham appears to have originated in 1922, with Alderman W. A. Cadbury's opposition to the extension of the General Hospital. Cadbury opposed the plan on the grounds that it would be much better to arrange an extension at a more suitable site in the suburbs than in the city centre.[129] Though many others initially opposed the suggestion for a number of reasons and an extension on the St Mary's site was permitted to proceed, a scheme to build such a centre in Bournbrook gradually gained support. By April 1930, an appeal for the funds to build the hospital was launched in the Town Hall. By December 1931, cash and firm promises reached £625,000.[130] With all other appeals placed on hold, existing voluntary hospitals began to feel the financial strain of a new Hospitals Centre. The pressure on hospital beds, however, had been relieved given an increase in accommodation at the municipal hospitals, under local authority control since the introduction of the Local Government Act (1929). By removing the taint of the Poor Law, the demand on municipal hospitals increased and the problems of hospital co-ordination became only more pressing.[131] As a result, the need for a Hospitals Centre began to be seriously questioned in some quarters.[132] Resistance would continue, mainly on financial grounds, until the first bricks were laid at the new site.

Although the Queen's Hospital's extension was permitted to proceed, work at the General was immediately halted. Without any display of enthusiasm, both hospitals finally agreed to promote a Bill in Parliament to amalgamate the two charities.[133] By the time amalgamation took effect in 1933, with Sir Charles Hyde, proprietor of the *Birmingham Post* and *Birmingham Mail*, elected the first President of the United Hospital, opposition to the Centre gradually died down. Staff at Birmingham's smaller general hospital, however, remained anxious about their identity, requesting that the 'old spirit of the Queen's Hospital not be allowed to die'.[134] In turn, many traditions in hospital organisation were challenged by the new Centre, including the dominance of the costly pavilion system of layout,

[129] Barnes, *Birmingham Hospitals Centre*, 37.

[130] Ibid., 69.

[131] Pickstone, *Medicine and Industrial Society*, 272.

[132] BCLA, Queen's Hospital, Management Committee Minutes, 1925–31, HC/QU/1/1/9.

[133] Barnes, *Birmingham Hospitals Centre*, 59.

[134] BCLA, Queen's Hospital, Medical Committee Minute Book, 1932–7, HC/QU/1/5/1.

which was replaced with the corridor system, each block comprising three or more stories. Unlike the General Hospital, with its wards of 16feet high, those at the Hospitals Centre were a less lofty 12 feet, and the building relied on natural ventilation.[135]

By 1935, the Queen's and General hospitals began to issue a joint annual report. Given that the new hospital would not be ready until January 1939, the city's two main voluntary hospitals would continue to operate in unison for several years, treating more than 100,000 patients annually.[136] Opening ceremonies arranged for 4 October were cancelled due to the 'serious international crisis', an alternative function being postponed until December.[137] Eventually, the King and Queen visited the school, the Queen agreeing to a decision to name the hospital in her honour. Thereafter, the Hospitals Centre became the Queen Elizabeth Hospital. Intended to replace an institution named after another queen, the hospital's new title allayed some lingering concerns and helped perpetuate the legacy of the city's original purpose-built teaching hospital.

Though originally proposed as the new home of the Maternity Hospital, the Queen's Hospital was unsuited to a maternity service given its location.[138] It was, however, extremely well suited to the emerging field of traumatology, the longest lasting 'orphan' of orthopaedics.[139] Despite many important advances in orthopaedic work in the early twentieth century, a study by the British Medical Association in 1935, which suggested that 35 per cent of fracture patients were disabled for the remainder of their lives, further encouraged the development of specialist rehabilitation services.[140] War injuries would only aggravate these limitations.[141] So long treating many of the city's industrial injuries, the Queen's Hospital was, very appropriately, transformed into the Birmingham Accident Hospital in 1941. Modelled on Lorenz Böhler's 125-bed Accident Clinic (1925), which shared a building with the Austrian National Insurance Company, its chief financiers, the 120-bed hospital in Birmingham opened with support from local industrialists, the Medical Research Council and the Ministry of Labour.[142] Under the clinical direction of an Australian, William Gissane,

---

[135] Barnes, *Birmingham Hospitals Centre*, 90.

[136] BCLLS, Birmingham United Hospitals, Annual Report, 1939.

[137] UBSC, University of Birmingham, Faculty of Medicine Minutes, 1938–41.

[138] BCLLS, Birmingham United Hospital, Annual Report, 1938.

[139] LeVay, *History of Orthopaedics*, 145.

[140] Borsay, *Disability and Social Policy*, 57.

[141] R. M. Titmus, *Problems of Social Policy* (London, 1950), 476–7.

[142] BCLA, Birmingham Accident Hospital, Board of Management Minutes, 1941–8, HC/AH/1/1/1.

it operated on three basic principles, emphasising the separation of the ill from the injured, continuity of care, and rehabilitation. Along with the Stoke Mandeville Centre for Spinal Injuries in Aylesbury, the Birmingham Accident Hospital became one of the nation's two specialised trauma centres;[143] it remained in the fore of accident medicine by further advocating prevention. Within a generation, the hospital's traditions, feared lost through reorganisation, were 'rebuilt in unified form',[144] and the century-old hospital had become world famous.

[143] LeVay, *History of Orthopaedics*, 145.
[144] Barnes, *Birmingham Hospitals Centre*, 59.

CHAPTER 10

Modernising Medical Education
in the Midlands

JUST AS LOCAL hospital administrators began to express a need for additional beds following the First World War, governors at Birmingham's medical school were equally concerned with clinical matters in those post-conflict years. Beginning with maternity beds, the medical faculty attempted to make up the most obvious deficiencies in clinical instruction during these years. Generally, this involved improving links between the medical school and the voluntary hospitals. However, in some cases, bed shortages encouraged medical educators, albeit reluctantly, to make better use of municipal hospitals.

By 1930, the medical specialties represented by local voluntary hospitals, from paediatrics to psychiatry, established a more clear presence in the medical school curriculum. Although usually associated with the nineteenth century, specialisation in medicine gained considerable impetus in the first decade of the twentieth century, a process aided by the First World War. As a consequence, the number of specialist medical institutions nationally nearly doubled between 1911 and 1921 alone.[1] In subsequent years, consultants and specialists blended together and became virtually interchangeable, both terms having been defined by the possession of hospital posts and distinguishing them from general practitioners.[2] At the same time, students appeared to be spending ever greater proportions of their time travelling between hospitals, as well as classrooms, with academic departments spread across two sites, one in the city centre, the other 2 miles distant in Bournbrook, Edgbaston. While both students and staff were spending many more hours in laboratories, which gradually transformed from side rooms into show rooms, members of the medical faculty travelled as much if not more, between sites, undertaking private practice, as well as their teaching and hospital duties. As a result, the problems facing medical educators in Birmingham only really began to be solved following a decision to consolidate medical and educational work at a single site in Edgbaston. Transforming part-time teaching appointments into full-time academic posts was an even greater challenge and continued into the 1940s.

---

[1] Pinker, *English Hospital Statistics*, 61–2.
[2] Weisz, *Divide and Conquer*, 183.

BEDS AND BODIES:
IMPROVING CLINICAL INSTRUCTION

During the interwar period, the changing nature of medicine forced medical instructors to rethink clinical instruction and provide greater training in emerging specialties. Ironically, it was only in the years following the First World War, when the maternal mortality rate in England was in decline, that governors first addressed the shortage of maternity cases available to students.[3] Specialist municipal provision in this area, on the other hand, had sharply increased following the passage of the Maternity and Child Welfare Act of 1918. In the following two decades, the medical profession became ever more committed to the promotion of hospital-based births, as is indicated in a steady rise in institutional births. In 1927, approximately 15 per cent of all live births took place in hospitals and other medical institutions; in the late 1930s, this had risen to 35 per cent. During the same period, the number of infant welfare centres more than doubled, as had the number of health visitors since the Act's passage in 1918.[4]

With governors at the General Hospital threatening to close fifty beds due to a lack of funds, the faculty members at the medical school looked to the reformed municipal health services for help. Appearing to have acted on advice given by Lawson Tait decades earlier, the medical faculty recognised the former workhouse infirmary, renamed Dudley Road Hospital in 1921, as a teaching hospital in order to access thirty-five obstetric and thirty-two gynaecological beds in the institution's Special Department for Midwifery and Diseases of Women.[5] Although Lawson Tait had suggested as early as 1890 that the recently completed infirmary on Dudley Road, with space for 1,500 patients, be transformed into the backbone of medical education in Birmingham, his views would require considerable time to catch on.[6] Though Professor Thomas Wilson was appointed consulting obstetrician to the hospital and a suitable laboratory with a technical assistant was immediately established, time spent in the maternity wards at the Dudley Hospital was not regarded as equal to attendance at other associated teaching hospitals. For the duration of this particular arrangement, two months at Dudley Road were valued as equal to one month's clerking at the Queen's Hospital. While a scheme formally to link the two main

---

[3] I. Loudon, 'On maternal and infant mortality, 1900–60', *Social History of Medicine* 4 (1991), 42.

[4] J. Lewis, *The Politics of Motherhood: Child and Maternal Welfare in England, 1900–39* (London, 1980), 117–39; Harris, *Origins of the British Welfare State*, 232.

[5] UBSC, University of Birmingham Medical School, Faculty Minute Book, 1921–24.

[6] Hearn, *Dudley Road Hospital*, 18.

teaching hospitals in 1921 would have resolved any further bed shortages, this particular solution was discouraged by hospital governors and medical staff, who preferred to establish an external midwifery service of their own in 1922.[7]

Given the perceived status of the municipal hospital system, relations between Birmingham's two hospital systems remained uncertain for the duration of this period. Throughout the 1930s, 'occasional perspicacious' students continued to visit Dudley Road for the purpose of making rounds, but only unofficially.[8] Relations with the hospital's pathology department were somewhat better, William Whitelaw, the consultant pathologist, having been a university examiner. Despite these links, as long as the consultants at Dudley Road were not Birmingham graduates, relations between the hospital and medical school remained cool. As a result, clinical instruction took many more years to develop along the lines suggested by Tait. Only in 1974 did students begin officially to receive instruction in medicine and surgery at the institution.[9]

Clinical work on venereal disease, the incidence of which escalated during the war, also required reorganisation in these years, especially as the Skin Hospital no longer treated the same number of venereal patients as in former years. Additionally, in the years immediately following the First World War, its facilities appeared out of date and venereal clinics opened at other institutions. As a result, links between the hospital and the medical school progressively weaken during these years. In response, students began to observe and treat venereal cases at the General and Queen's hospitals, the latter institution opening its own venereal department in 1918.[10] In total, hospital staff gave eight clinical lectures, male students attending two on syphilis and two on gonorrhoea at each institution, female students receiving only four lectures in total.[11] Moreover, the addition of a twenty-six-bed venereal clinic at Dudley Road, one of 190 treatment centres set up throughout the country between 1917 and 1920,[12] further increased ties between the former Poor Law infirmary and the medical school. Clearly the less desirable of solutions, these links also weakened in 1930, when the Skin Hospital was again recognised as an associate teaching hospital.

Another area that demanded the attention of the medical faculty was psychiatric medicine. Interestingly, the staff at the medical school was

7 BCLA, Queen's Hospital, General Committee Minutes, 1917–25, HC/QU/1/1/8.

8 Hearn, *Dudley Road Hospital*, 48.

9 Ibid., 103.

10 BCLA, General Hospital, Medical Committee Minute Book, 1916–21, GHB 77.

11 UBSC, University of Birmingham, Faculty of Medicine Minutes, 1921–4.

12 Hardy, *Health and Medicine in Britain*, 69.

slower to respond to developments in this field, psychiatry, like orthopaedics, having forged its place as a key specialty during and immediately after the First World War. Alongside greater interest in shellshock and psychoanalysis, new research emphasised a closer relationship between physical and mental illness. While the period witnessed a more optimistic approach to mental health, psychiatric institutions became even more central to mental health care. These changes are also reflected in the introduction of the Mental Deficiency Act in 1913, which aimed at the compulsory institutionalisation and segregation of the feeble-minded, whose numbers were to have risen substantially between 1913 and the late 1930s. Both the 1913 Act and its successor, the Mental Deficiency Act of 1927, promoted institutional approaches to the treatment of mental deficiency.[13] In Birmingham, further interest in the subject appears to have been inspired by the decision in 1919 of Sir Charles Hyde, proprietor of the *Birmingham Post*, to leave the school with £2,000 with which to found a Lectureship in Psychotherapy.[14]

The inaugural lecture, on the 'Principles of Psychotherapy', was given by William McDougall in 1920, the same year the Queen's Hospital opened its Department of Psychotherapy. McDougall, the first experimental psychologist appointed to Oxford, went on to formulate the first theory of human instinctual behaviour. Though Hyde originally intended the lectureship to benefit both staff and students, the first lectures were not open to students, who presented the faculty with a collective wish to attend a subsequent series of ten postgraduate lectures in 1922 given by Frederick Mott, cofounder of the Maudsley Hospital in London. Despite undertaking what has been described as the most important stream of psychiatric research in the country before the First World War, Mott's lectures were described as 'studentish' by local practitioners who attended the event, and the lectures, renamed the Lectureship in Morbid Psychology, were thereafter open to students.[15] Staff also considered the time ripe for more instruction in psychoanalytic theory. A chair of Psychiatry, on the other hand, would not be appointed in Birmingham until 1962.[16]

In the post-war period, clinical instruction in mental diseases was divided between Dr Percy Hughes and Dr Charles Forsyth; students were given the option to observe patients at Worcestershire County Mental Hospital, Barnsley Hall, Bromsgrove, or the City Hospital at

[13] Harris, *Origins of the British Welfare State*, 233.
[14] UBSC, Particulars of Postgraduate Lectures and Demonstrations, Faculty of Medicine, 1877–1939.
[15] UBSC, University of Birmingham, Faculty of Medicine Minutes, 1921–4.
[16] [Birmingham Regional Hospital Board], *Birmingham Regional Hospital Board*, 70.

Winson Green. Additional 'illustrations of living examples' became available to students following the affiliation of the Queen's Hospital with the Birmingham Nerve Hospital in 1923, though the former institution tended to treat neurological outpatients only.[17] Founded in 1914, the Nerve Hospital was located at 20 Bath Row, which remained the institution's outpatient department until 1948. The Queen's Hospital eventually opened its own mental health outpatient clinic following the introduction of the Mental Treatment Act (1930), which made provisions for voluntary admission to psychiatric hospitals.[18] At this later date, medical instruction also appears to have been more regularly influenced by the introduction of mental health legislation, including an interdepartmental report on mental deficiency in 1929 (the Wood Report), which gave advocates of eugenic policies a second wind. While arguments in favour of sterilisation appeared to be gaining support across the country, staff at the medical school introduced additional instruction on mental deficiency at this time.[19]

Another area of clinical instruction regarded as deficient was access to patients suffering from pulmonary tuberculosis. Though the number of such cases nationally had been in decline since 1880, TB remained the most visible of chronic diseases affecting men and women in the most productive years of their lives, at least statistically.[20] Moreover, the rise of tuberculosis among young women during the war, and the return of some 58,000 tuberculous ex-servicemen in the immediate aftermath of the conflict, justified renewed interest in the disease.[21] With the establishment of a national tuberculosis service in 1913, however, many more cases were transferred from urban hospitals to rural sanatoria. As a result, in Birmingham, and in many other provincial cities, students encountered very few cases of TB on the wards of local teaching hospitals.

In 1922, the medical faculty at Birmingham attempted to remedy this shortage. Together with the city's TB service, staff arranged for cases to be passed to hospital outpatient departments where they would be treated by students before being sent on to sanatoria. In their fourth years, students gained access to Yardley Road Sanatoria where they would observe institutional treatment under the direction of the City TB officer three Saturdays in the summer.[22] Despite these arrangements, further instruction was required, the Secretary of the Conjoint Examining Board in London

---

[17] BCLA, Queen's Hospital, General Committee Minutes, 1917–25, HC/QU/1/1/8.

[18] BCLA, Queen's Hospital, General Committee Minutes, 1925–31, HC/QU/1/1/9.

[19] UBSC, University of Birmingham, Faculty of Medicine Minutes, 1925–9.

[20] Hardy, *Health and Medicine in Britain*, 53.

[21] Ibid., 85.

[22] UBSC, University of Birmingham, Faculty of Medicine Minutes, 1921–4.

conveying by letter the following June that their examiners were 'unfavourably impressed by the frequent inability of [Birmingham] candidates, otherwise fairly well informed, to recognise the physical signs of ordinary pulmonary disease even when present in a marked degree'.[23] In subsequent years, students received ten lectures on TB, in addition to the occasional clinical lecture. Furthermore, during the years 1929 to 1933 alone, 234 cases of pulmonary TB were admitted to the General for teaching purposes, cases representing every stage of the disease and providing opportunities to observe open-air and rest treatment, the use of tuberculin, sanoctysin and artificial pneumothorax, as well as surgical methods, such as phrenic avulsion and thoracoplasty.[24] Despite these improvements, at the General Medical Council's congress in 1934, only members of the Birmingham school continued to report an occasional shortage of TB cases for teaching purposes.[25]

In general, these and other episodes demonstrated to the medical faculty that a closer relationship between the school and the teaching hospitals was desirable. An increase in bed numbers was also suggested. Consequently, in November 1922, a board was formed to consider both issues. By 1925, the answer no longer appeared to depend on improved communication between the medical school and local hospitals, than with plans to enlarge the General Hospital in order that medical educators could be 'kept abreast of modern requirements, and opportunities to see all types of cases be afforded the students'.[26] The following year, plans to enlarge Birmingham's oldest voluntary hospital soon gave way to the decision to build a new Hospital Centre in order to raise the numbers of local hospital beds to 2,400 by 1931 (see Chapter 9).

MEDICINE'S PAST AND MEDICINE'S FUTURE

Although the school's centenary was drawing near, with so much change occurring in these years, the time appeared ripe for staff to begin to reflect on the history of medical education in Birmingham. In November 1924, members of the medical faculty began to plan their centenary in 1928. By May 1925, however, minutes reflect opinion over the centenary to have been divided, as continues to be the case today,[27] with J. T. J. Morrison, a Professor of Forensic Medicine, tracing the school's foundation to 1

---

[23] Ibid.

[24] Ibid.

[25] Ibid.

[26] UBSC, University of Birmingham, Faculty of Medicine Minutes, 1924–5.

[27] J. Reinarz, 'Unearthing and dissecting the records of English provincial medical education, *c.* 1825–1948', *Social History of Medicine* 21 (2008), 388.

December 1825, when Sands Cox commenced lectures in his home and practice, and not 1828 when an actual school was established at Snow Hill. With staff at Sheffield tracing the foundation of their school to July 1828, the Birmingham faculty not surprisingly decided on the earlier date, making it the earliest provincial medical school after Manchester. That December, members of the faculty marked their centenary with a special dinner at which four honorary degrees were conferred, an address was given and a conversazione held. While Morrison went on to write a comprehensive history of the school which celebrated its founder, other memorials to Cox were considered, including a lectureship in connection with medical history, ethics and sociology, as well as naming the departments of anatomy and physiology after the school's founder.[28] Instead, in January 1926, the faculty renamed the postgraduate lectureship in morbid psychology (founded by Charles Hyde) the Sands Cox Memorial Lectureship. Unfortunately for the school's unsung founder, only two months later it was decided to rename the postgraduate lectureship after William Withering, whose name it was believed would lend the school greater prestige. Appropriate to this dispute involving two of the most important historical figures in Birmingham medicine, Charles Singer, one of the first English historians of medicine, was invited to deliver the Withering lecture the following year.[29]

Given such changes to postgraduate lectures and a hesitancy to rename an entire department in the surgeon's honour, the £21 Sands Cox Scholarship remained one of the sole references to the school's founder. By the late 1920s, it was one of fifteen entrance scholarships available to the approximately sixty medical and thirty dental students admitted to the school annually. In general, the number of students entering the school each year had stabilised before staff decided to limit the number of new students to sixty in December 1931. In total, student numbers actually dropped between 1921 and 1931 from 445 to 338, though women continued to comprise approximately a quarter of new entrants. Between 1900 and 1936, a total of 279 women trained at Birmingham medical school.[30] The school was also attracting increasing numbers of foreign students, who tended to be older than local students. Consequently, students were also described as being more mature than in the past. Unfortunately, their academic performance had hardly improved since the first decade of the twentieth century.

Contrary to expectations, examination results continued to suffer in

---

[28] UBSC, University of Birmingham, Faculty of Medicine Minutes, 1924–5.

[29] UBSC, University of Birmingham, Faculty of Medicine Minutes, 1925–9.

[30] *Birmingham Post*, 12 October 1936.

the interwar years. For example, the Conjoint Board examinations of the two royal colleges in 1923 were a great disappointment to staff, with the number of students failing the exam being nearly double the national average.[31] A third of local students sitting the final MB in 1924 also failed. In March 1925, Birmingham's rejections at examinations were higher than the national average in all subjects, except medicine and midwifery. The highest rejection rates were in anatomy and physiology, with more than 50 per cent of students failing their examinations. Desperate for an explanation, faculty members blamed the poor performance on the school's foreign students, many of whom had difficulties communicating in English. The number of Egyptian and Asian students in particular was said to have necessitated the introduction of English classes in 1924.[32] Neither did the number of American students in regular attendance improve results. According to the GMC, the numbers of American students coming to Britain was 'most undesirable', with the worst usually applying to English schools.[33] With 262 Americans applying in 1931 alone, their numbers might easily have been increased in the years before the Second World War. Instead, their numbers were restricted to three a year, while students from Egypt were limited to two a year. Generally, the school privileged only those foreign students who intended to sit for university degrees. Those with unacceptable examination results, on the other hand, were encouraged to abandon their studies entirely.

Other changes, especially those intended to improve the social lives of students, are more likely to have had a positive effect on academic performance. For example, in 1924, after receiving a letter from the Medical Students' Society, whose members requested the introduction of a free afternoon a week, staff commenced arrangements for a Wednesday half-holiday. By 1928, the half-holiday had become reality.[34] Consequently, student societies and recreational activities flourished, while those individuals desirous of additional time to catch up on reading and laboratory work found themselves with a free afternoon. While Reverend Berry perhaps expected better attendance at the religious services he arranged for medical students at Queen's College since 1923, things had changed substantially since Reverend Warneford's day.[35] Though the subject of 'hot debate' in these years, religious training by 1934 ceased to be regarded as part of a doctor's education, even in the eyes of the General Medical

---

[31] UBSC, University of Birmingham, Faculty of Medicine Minutes, 1921–4.

[32] Ibid.

[33] UBSC, University of Birmingham, Faculty of Medicine Minutes, 1929–32.

[34] UBSC, University of Birmingham, Faculty of Medicine Minutes, 1925–9.

[35] UBSC, University of Birmingham, Faculty of Medicine Minutes, 1921–4.

Council.[36] There were of course those who continued to regard medical students as debauched and disorderly, though faculty members appear to have considered alternative explanations for underachievement. For example, student performance might as easily have suffered given the age of the school, Professor Whitehouse having complained to staff about the discomfort of sitting in the lecture theatre in 1932.[37] While the construction of a new medical school would remedy such complaints, and perhaps improve concentration, other less complicated ways of enhancing student performance were also implemented.

In the 1930s, results appear to have improved in some subjects, with students in anatomy, physiology, pharmacology, medicine and midwifery performing better than the national average. There was still room for improvement in chemistry and surgery, with rejection rates reaching 75 per cent and 62 per cent, compared with national averages of 43 per cent and 48 per cent respectively.[38] In 1934, staff suggested further improvements were near, with claims that the type of student who kept exam rejections high would be gone in a year or two.[39] Keeping to their promise, staff advised four students to withdraw the following year on the basis of their recent examination results. Further improvements came in 1929 with the introduction of a sub-dean, who scrutinised the backgrounds of candidates more carefully, as well as new methods of student selection in October 1937. The apparently informal and *ad hoc* admissions procedures at medical school in the interwar period has recently been construed as an example of the 'old boy system in action', though clearly selection, at a deeper level than social networking, was occurring at a number of schools in these years.[40] At Birmingham, selection alone appears the most convincing explanation for the success of students at medical examinations in these years.

STANDARD, YET SPECIALISED: CURRICULUM CHANGE IN THE 1920S AND 30S

While teaching methods at medical schools in these years are difficult to assess, changes to the curriculum are apparent, with many modifications doubtlessly encouraging better examination results. While some subjects, such as materia medica and forensic medicine, were declining in

[36] *Birmingham Post*, 30 May 1934.

[37] UBSC, University of Birmingham, Faculty of Medicine Minutes, 1929–32.

[38] Ibid.

[39] UBSC, University of Birmingham, Faculty of Medicine Minutes, 1933–6.

[40] K. O'Flynn, ' "Intellectual athletes": changing selection criteria for London medical schools from 1918–1939 to 2002', *Journal of History in Higher Education* 73 (2004), 11–23.

importance, others, such as orthopaedics, the diseases of children, as well as those of the ear, nose and throat, were more regularly the subject of lectures. While the latter is hardly surprising given the numbers of children undergoing surgical removal of adenoids and tonsils at local and national clinics, paediatrics was transforming rapidly in these years. For example, while Leonard Parsons had lectured twice a week each summer since 1914 on infant hygiene and diseases peculiar to children, in 1925 he was requested to report on child medicine as taught at other medical schools with the aim of introducing some reforms.[41] While lectureships in orthopaedics and diseases of the ear, nose and throat were also supported, albeit as options in the fifth year, Parsons's report convinced the committee that additional teaching in childhood disease was required. Significantly, the following year, the faculty selected George Still, the 'father of British clinical paediatrics',[42] as its Ingleby lecturer, though he was unable to travel to Birmingham on the occasion due to disruptions caused by the General Strike.[43]

In contrast, Still's discipline advanced unimpeded. By 1928, Birmingham had its own professor of paediatrics, Parsons having been appointed to the first Chair of Infant Hygiene and Diseases of Children. The same year witnessed the foundation of the British Paediatric Association, of which Parsons, along with Still, was a founding member.[44] Parsons was also instrumental in founding the Birmingham Institute of Child Health two years later, with a view to integrating efforts to prevent as well as cure the problems of childhood.[45] Meanwhile, the Children's Hospital was recognised as the Department of Diseases in Children, receiving in turn a grant of £50 for books to establish a hospital library to support teaching. Lectures, which were increased from ten to twenty-six annually, were transferred to the hospital and every student was required to serve as a clinical clerk in the hospital for four to six weeks. By 1932, a glimpse at the school's examination papers reveals that the place of childhood medicine had been secured, with 25 per cent of questions dealing with diseases of children.[46] Parsons's textbook *Diseases of Infancy and Childhood* (1933), written in collaboration with Seymour Barling, became a standard work in British paediatrics. The first member of the Birmingham faculty to deliver the Withering Lecture, Parsons in his work emphasised the importance of

[41] UBSC, University of Birmingham, Faculty of Medicine Minutes, 1925–9.

[42] Lomax, *Small and Special*, 164.

[43] UBSC, University of Birmingham, Faculty of Medicine Minutes, 1925–9.

[44] P. Dunn, 'Sir Leonard Parsons of Birmingham (1879–1950) and antenatal paediatrics', *Archives of Disease in Childhood* 86, 1 (2002), 65–7.

[45] Ibid.

[46] UBSC, University of Birmingham, Faculty of Medicine Minutes, 1929–32.

the experimental method and laboratory research in clinical medicine.[47] Under his direction, the Children's Hospital had included research results in their annual reports since 1926. Amongst numerous important studies undertaken by staff in these years, early reports highlighted research on rheumatism, rickets and celiac disease.[48]

In terms of its relatively young age, the place of radiology in medical instruction was secured more quickly than was paediatrics. Approximately two decades after the first x-ray departments were established at English voluntary hospitals, the British Association for the Advancement of Radiology and Physiotherapy had been set up to sponsor a new Diploma in Medical Radiology and Electrology at the University of Cambridge.[49] The Association's first examinations were held in 1920, the same year a Society of Radiographers was founded in London. Three years later, members of the medical faculty in Birmingham similarly expressed a desire to organise a course in radiology. Though students received clinical instruction in radiology as early as 1916,[50] a Lecturer in Radiology was appointed in 1924 to address the second-year students four times each summer. Commencing with the way in which x-rays were produced, lectures covered aspects of bone disease, the location of foreign bodies, examination of the respiratory, cardiovascular and alimentary systems, as well as treatment by radiation.[51]

Many other subjects originally entered the curriculum as postgraduate lectures, with sessions in 1921 covering medical penology (or prison medicine) and psychoanalysis, the latter proving particularly popular with several 'unqualified ladies', as well as students.[52] Additionally, postgraduate demonstrations were held each year at the school's two main teaching hospitals, events which usually comprised two dozen lectures, attracted the interest of several dozen local medical practitioners and grew more intensive after 1927, with lectures in dentistry commencing that same year. Finally, many of these subjects fuelled the research efforts of staff in these years and, in the face of a national curriculum, permitted schools to develop their own very distinctive, local profiles.

Just as some practitioners preached the benefits of radiological techniques, others advocated the advantages of more traditional remedies, such

---

[47] *Birmingham Post*, 14 May 1937.

[48] BCLA, Children's Hospital, Annual Report, 1927.

[49] Stevens, *Medical Practice in Modern England*, 46; Weatherall, *Gentlemen, Scientists and Doctors*, 202–3.

[50] UBSC, Faculty of Medicine, Minutes of the University Clinical Board, 1911–26.

[51] UBSC, University of Birmingham, Faculty of Medicine, General Information, Regulations and Syllabus, 1925–6.

[52] UBSC, University of Birmingham, Faculty of Medicine Minutes, 1921–4.

as the healing powers of water. In the years following the First World War, hydrotherapy witnessed a revival, with many soldiers and rheumatic patients attending scientifically run spas.[53] Standards at British spas continued to improve with the foundation of the British Spa Federation in 1921. Although historians have suggested that no courses in the subject were run at British universities,[54] following a communication from the International Society of Medical Hydrologists, a series of postgraduate lectures addressing spa therapies was commenced at Birmingham in 1924. The first of these specialist talks was delivered by Dr Rupert Waterhouse, physician to the Royal Mineral Water Hospital at Bath, and attracted an audience of forty.[55] Spa treatment had long attracted the interests of local medical practitioners, both John Ash and Arthur Foxwell having published on the subject in previous centuries.[56] Additionally, such therapy remained popular for many more years, especially as the cost of travelling abroad increased with a departure from the Gold Standard in 1931. The subject's acceptance by Birmingham's medical practitioners is perhaps best demonstrated by a decision to include a Department of Hydrology in the new Hospitals Centre when it opened.[57]

While Birmingham's social hydrologists appeared to challenge the secondary status of spa therapy, postgraduate teaching in other areas appeared better suited to the medical school and the community which its faculty served. The best example of this was the decision to develop a course on industrial hygiene, a subject particularly well suited to a medical school based in 'the first manufacturing town in the world'.[58] Throughout his career as chief factory inspector, Thomas Legge laid stress on the need for better education in occupational medicine at medical schools. In 1928, Legge was provided with an opportunity to address this omission when he was invited to Birmingham as that year's William Withering lecturer. Besides being well attended, Legge's lectures encouraged the medical faculty to organise its own postgraduate course on industrial medicine.

53 D. Cantor, 'The contradictions of specialisation: rheumatism and the decline of the spa in inter-war Britain', in *The Medical History of Waters and Spas*, ed. R. Porter (London, 1990), 127–31.

54 Cantor, 'The contradictions of specialisation', 137.

55 UBSC, Particulars of Postgraduate Lectures and Demonstrations, Faculty of Medicine, 1877–1944.

56 J. Ash, *Experiments and Observations, to Investigate, by Chemical Analysis, the Medicinal Properties of the Mineral Waters of Spa and Aix-La-Chappelle, in Germany; and of the Waters and Boue near St. Amand in French Flanders* (London, 1788); 'Arthur Foxwell', in *Oxford Dictionary of National Biography*.

57 BCLA, Queen's Hospital, Medical Committee Minutes, 1937–41, HC/QU/1/5/2.

58 Hopkins, *Rise of the Manufacturing Town*, xiii.

In 1933, school staff canvassed prominent local businesses and proposed the establishment of such a department at the new medical school. By June 1934, Donald Stewart, medical officer to Imperial Chemical Industries was offered the newly created Readership in Industrial Hygiene and Medicine, along with a £750 salary. Perhaps regarding the income as insufficient, Stewart turned down the post, accepting instead that of medical officer at Austin's at Longbridge, Birmingham, where he became responsible for the health of the automobile manufacturer's 18,000 employees. The readership was subsequently offered to and accepted by Howard Collier, a Quaker doctor from Redditch. Having settled in the Midlands in 1919 at the conclusion of the First World War, the Edinburgh-educated Collier, unlike Stewart, brought little industry, but much industrial experience to the post.[59] Before his appointment, Collier had worked as a certifying factory surgeon, medical officer of health and honorary surgeon to Smallwood Hospital, Redditch.[60] While in Redditch he also developed an interest in unorthodox systems of healing.[61]

By May the following year, Collier set about publicising the new department by delivering a series of postgraduate lectures, entitled 'Preventive Medicine in Industry'.[62] With the support of an advisory board containing representatives of various employers' associations, chambers of commerce, insurance companies, as well as the Birmingham Trades Council, Trades Union Congress and Home Office, Collier rapidly developed the department's teaching and research profile. The first research subjects he proposed concentrated on the dermatitis-producing substances in turpentine, the effect of titanium oxide on the lungs, the carcinogenicity of pitch-containing paint sprays, as well as the effect of paraffin wax and silicate dusts on the lungs.[63] From 1935, the department's work was assisted by Dr Ester Killick, formerly of the Mines Research Department, whose particular interests lay in carbon monoxide poisoning. Like Collier, Killick believed the best way to detect the effects of poisons was to try them on human subjects. As a result, her research involved spending many hours at the medical school in a sealed box breathing carbon monoxide.[64] Equally important to the department's work was a museum of industrial hygiene, which included specimens of an industrial interest, including x-rays, protective devices and a large statistical collection. Attended by doctors and

[59] H. E. Collier, *Experiment with a Life* (Wallingford, PA, 1953), 6.

[60] *Birmingham Post*, 8 November 1934.

[61] Collier, *Experiment with a Life*, 7.

[62] *Birmingham Post*, 2 May 1935.

[63] UBSC, University of Birmingham, Faculty of Medicine Minutes, 1932–6.

[64] *Birmingham Daily Mail*, 17 June 1936.

industrial nurses, Collier's courses soon included one on industrial medical psychology and another on industrial medical organisation. Costing 6 and 7 guineas respectively, courses ran for two weeks and comprised twenty lectures, six clinical demonstrations, as well as factory visits. His course on industrial medicine covered subjects from eye damage to blood examinations for industrial toxaemias, while a second course was designed to help medical officers appoint workers to those tasks that suited them best.[65]

After only two years, Collier's department appeared a success, signalled, among other things, by his appointment to the International Committee on Industrial Medicine (Geneva). In 1936, he developed the clinical side of his department by establishing a consultative clinic, to which patients were referred to by GPs, industrial medical officers or manufacturers. Acting as an independent authority, strictly guarding the anonymity of workers, the clinic's staff, which included a physician, haematologist, dermatologist and industrial hygienist, researched the effects of acetone, carbon-monoxide poisoning, lead and cyanide dust.[66] A survey of 15,000 workers in the region in 1936 demonstrated, among other things, that overtime caused excessive fatigue, as did workers' longer journeys to and from their workplaces.[67] A further comprehensive report issued by the department in 1937 suggests Collier was active in devising a standard method of sickness reporting at Midlands firms, in order to aid the collection of accurate and useful industry statistics. He also investigated deaths from chronic rheumatitis, arthritis and gout, while the consultation centre at the General Hospital helped reveal cases of nose ulcers caused by faulty ventilation, sepsis among metal japanners, ten cases of skin trouble caused by the incorrect use of a new solvent, as well as dermatitis among local hat sizers.[68] Besides contacts with industrial research committees nationally, Collier also commenced a series of 'staff-interchanges' between Birmingham and Harvard, where a similarly funded division of industrial medicine had existed in the Department of Public Health since 1918.[69]

Despite such innovative and collaborative work in the field of occupational health and preventive medicine, Collier's department, like his research subjects, was in poor health. Given the political and economic climate in the late 1930s, financial support for the department soon diminished. By February 1939, the faculty declared that, unless the department

[65] UBSC, University of Birmingham, Faculty of Medicine Minutes, 1932–6; 1936–8.

[66] UBSC, University of Birmingham, Faculty of Medicine Minutes, 1932–6.

[67] *Birmingham Post*, 12 October 1936.

[68] UBSC, University of Birmingham, Faculty of Medicine Minutes, 1936–8.

[69] UBSC, University of Birmingham, Faculty of Medicine Minutes, 1938–41.

of industrial hygiene could appoint a research worker for £1,000 and for about ten years, its work would cease.[70] Though governors of voluntary hospitals had issued similar claims on many previous occasions, this was no empty warning. By May, with no further funds forthcoming, the department's closure was announced. Its contribution to local medical practice, however, was clearly recognised. By February 1941, it was again suggested that a course in industrial medicine be offered and the department be reopened. In the following years, similar departments emerged in Manchester, Glasgow and Newcastle. During the post-war period, these departments, as well as the more famous Department of Research in Industrial Medicine, established at the London Hospital in 1943 and run by Donald Hunter, attracted not only more funding than Birmingham, but also the attention of historians.[71] Unlike Hunter's role in promoting industrial hygiene research and education, Birmingham's has been largely forgotten.

INVESTIGATE AND RADIATE:
PROMOTING RESEARCH IN THE INTERWAR YEARS

In contrast to the earliest centres for industrial medicine, physiology departments, described as the model medical science, have more often attracted the attention of historians, given their apparent fondness of laboratories. With the continued growth of experimental physiology in the twentieth century, medical students spent as much time in laboratories as they did in lectures. While it has been argued that the medical sciences, such as physiology, were promoted most actively in the provincial schools,[72] such developments appear to have taken some time to materialise in Birmingham. By 1925, laboratories began to proliferate at both the medical school and voluntary hospitals. In that same year, the white lab coat had become compulsory attire for all students attending the General Hospital.[73] The growth of experimental physiology commenced the following year when additional space became available at the medical school following the Zoology Department's move from the Edmund Street campus to Edgbaston. Staff immediately set about building accommodation for animals used in the department's research. The following year, 1926, the department was reorganised at a cost of £2,500, its staff doubling in size,

[70] Ibid.

[71] J. A. Bonnell, 'Donald Hunter', *Journal of the Society of Occupational Medicine* 29, 2 (1979), 81; R. Schilling, 'Donald Hunter, 1898–1978: editor BJIM 1944–50', *British Journal of Industrial Medicine* 50, 1 (1993), 5–6.

[72] Waddington, *Medical Education at St Bartholomew's*, 116.

[73] BCLA, General Hospital, Medical Committee Minute Book, 1925–34, GHB 80.

appointments including a lecturer in biochemistry, additional assistant lecturers and a boy to attend the animals.[74]

While these changes appear to have improved practical instruction in physiology, the research work of the department also began to flourish. Instead of designing its own distinct research agenda, however, the Department of Physiology followed an agenda of national cancer research. Public concern about cancer reached a new peak in 1923, when the Ministry of Health noted that the mortality rate from the disease had risen 20 per cent between 1901 and 1921.[75] Rising cancer mortality also stimulated interest in radium as an alternative to surgery, which often promised long and painful operations and possible disfigurement. At £14,500 per gram, however, the element's cost inhibited its wide use. Consequently, the Medical Research Council acquired 5 grams of hydrated radium bromide from surplus government stock for the basis of their radiological research programme. In 1921, the council established its Radiological Committee, while the Ministry of Health set up a Cancer Advisory Committee a year later. A philanthropic research organisation, the British Empire Cancer Campaign (BECC), founded in 1923, also improved access to radium. The following years witnessed a 'radium boom', and many practitioners, formerly critical of radium therapy, began to praise its benefits.[76]

Having received 290 milligrams of radium bromide from the Medical Research Council in 1920, the Department of Physiology at Birmingham began to develop its own programme of cancer research. The majority of radium, however, was used for treatment, not research. Five years later, the board of the General Hospital was empowered to appoint two representatives to the local committee of the BECC and set aside two beds for cancer research. Another twelve were set aside at the Jaffray Hospital, where staff focused their attention on carcinoma of the rectum and oesophagus.[77] In return, the BECC soon lent the institution a further 270 milligrams of radium. The medical school's Cancer Research Scheme was based at the Jaffray, where twelve beds had been placed under the management of Professor F. W. M. Lamb in 1926. In December 1928, research was supported through the establishment of a laboratory specifically designed for the investigation of tissue culture under the leadership of Dr H. Ines Pfister in the Department of Physiology. Following visits to similar labs

[74] UBSC, University of Birmingham, Faculty of Medicine Minutes, 1925–9.

[75] D. Cantor, 'The MRC's support for experimental radiology during the inter-war years', in *Historical Perspectives on the Role of the MRC: Essays in the History of the Medical Research Council of the United Kingdom and its Predecessor, the Medical Research Committee, 1913–1953*, ed. J. Austoker and L. Bryder (Oxford, 1989), 185.

[76] Ibid., 189.

[77] BCLA, General Hospital, Medical Committee Minute Book, 1925–34, GHB 80.

in Berlin, Turin and Tokyo, Pfister used the lab to observe cell growth under the influence of x-rays, radiation and chemical agents, recording growth with a cinematograph.[78] While the laboratory was licensed for animal experimentation, in future, the MRC's radium was to be used for research only. For this reason, in 1928, the General Hospital purchased another 100 milligrams and planned to acquire 300 more. The following year, this increased to 600 milligrams. After this purchase, thirteen beds were set aside for radiation treatment in the hospital's old venereal block and another twenty beds were quickly set aside. Mr F. S. Phillips, who had been appointed radium curator, was also soon entrusted with another gram of radium when the National Radium Commission selected the hospital as a centre for treatment and research into the use of radium in cancer.

National radium supplies increased dramatically in the 1929–30 period. Allocation of the element for research purposes expanded with the efforts of the MRC and the newly formed Radium Trust and Commission.[79] By November 1929, cancer research at the school reached a new level following a decision by the National Radium Commission to establish a series of national radium centres. That in Birmingham was based at the medical school and its two main teaching hospitals. Under the scheme, cancer beds at the General increased to forty-three, and twenty-five at the Queen's Hospital were designated for similar research, and a lab under the direction of two lecturers in the Department of Physics would be used for the local production of radon.[80] The following year, staff at the General and Queen's hospitals, as well as twelve members of institutions outside Birmingham, attended a postgraduate course in radium therapy. By 1931, the radon lab had been completed and a full-time researcher was appointed to the cancer tissue laboratory. Over the next three years, 1,000 cancer cases would benefit from locally produced radon, demand of which had risen from 110 millicuries per month to 315, a fifth of the plant's capacity.[81] By 1933, it was noted that great improvements in the treatment of cancers of the mouth and larynx had been achieved and Professor Beckwith Whitehouse was made a member of the National Radium Commission. Finally, connections between radiology and the radium centre were made following the establishment of a deep x-ray therapy department, donated to the hospital by Sir Herbert Austin. Though raising nearly £200 from radon sales to hospitals and private patients, the cancer centre, despite conducting much important research, remained a great cost. Though production at Canadian

---

[78] UBSC, University of Birmingham, Faculty of Medicine Minutes, 1925–9.

[79] Cantor, 'The MRC's support for experimental radiology', 197.

[80] UBSC, University of Birmingham, Faculty of Medicine Minutes, 1929–32.

[81] UBSC, University of Birmingham, Faculty of Medicine Minutes, 1932–6.

pitchblende mines helped reduce the price of radium to £7,000 a gram in 1936, voluntary hospitals could not acquire enough to satisfy the demand. By this time, treatment involved exposing patients to larger quantities of radium, concentrated in a receptacle referred to as a 'bomb'. As a result, with 1,469 people in Birmingham dying of cancer in 1935 alone, more was urgently needed.[82]

Together with the carbon-monoxide work of the Industrial Hygiene Department, investigations into rheumatism at the Children's Hospital and tissue cultures at the Department of Physiology, the cancer work carried out at the General Hospital was regarded as among the most important research being undertaken in Birmingham in these years.[83] As short as this list appears, research work, especially on the wards of the local voluntary hospitals, was on the increase in the 1930s. Even routine work at local hospitals in the nineteenth century, however, had permitted the accumulation of data for research purposes. The same was true for the twentieth century. For example, given that the majority of the pathologist's work at the Women's Hospital involved the detection of cancerous tumours, over time the department's records became of great use to staff and researchers outside Birmingham working on ovarian cancer. The work of the Eye Hospital's pathologist for a time similarly concentrated on a single ailment, namely infantile ophthalmia.[84] At other institutions, however, the routine nature of clinical work very often hindered staff from undertaking any systematic research.

In order to promote research efforts in these years, a number of hospital boards established posts and set aside funds specifically designed to encourage research. For example, a Medical Faculty Research Fund was created in 1929, of which a £1,000 donation from Lord Beaverbrook formed the nucleus.[85] A year later, the estate of Miss Caroline Harrold was left to the university to fund research in science and medicine. By 1934, the Caroline Harrold Research Fund provided an additional £200 to £250 annually for research carried out at the medical school or its associated hospitals, where similar incentives were also being created.[86] For example, a Medical Progress Fund was set up at the Queen's Hospital in 1931 to support research and teaching.[87] Three years later, the hospital announced the creation of another research fund for medical graduates. A similar

[82] *Birmingham Post*, 4 April 1936.

[83] *Birmingham Post*, 6 October 1937.

[84] BCLLS, Eye Hospital, Annual Reports, 1922–31.

[85] UBSC, University of Birmingham, Medical Faculty Minutes, 1925–9.

[86] UBSC, University of Birmingham, Medical Faculty Minutes, 1932–6.

[87] BCLLS, Queen's Hospital, Annual Report, 1931.

fund was first suggested by the staff of the Eye Hospital in 1924, but was established only in 1936 following an initial donation of £750. While staff hoped another £10,000 might be raised, research work commenced with Dr Dorothy Campbell investigating sulphur metabolism in patients with senile and other forms of cataract.[88] At the General, a new clinical laboratory, dedicated to the memory of Dr James Russell, a former Professor of Medicine, opened in 1925. In 1928, staff advertised the creation of a Russell Research Studentship to help the hospital's biochemist, Garfield Thomas, collect specimens of gastric contents, Russell himself having originally collected many of the earliest specimens included in the pathology museum he helped found. Only a month later, Professor Wynn suggested the General Hospital establish an asthma clinic, an 'influential National Committee' having been formed to promote investigation into the disease and appeared willing to pay a part-time research worker in the event such a clinic was established.[89] In 1928, an anonymous donor presented the hospital with £250 annually to found a studentship in diabetic research under Dr Arthur Thomson. The scheme apparently flourished and, by 1933, the diabetic clinic reportedly attended 2,000 cases.[90] In 1930, funds from the Rockefeller Foundation promised even greater changes in the research culture of the medical school. With the Foundation's support, Professor Haswell Wilson, pathologist to the General, Queen's and Children's hospitals, accompanied by the dean of the medical school and the Professor of Physiology commenced a tour of American medical schools to examine the teaching of pathology.

Unfortunately, the department's research profile proved more resistant to change. As in previous years, the Department of Pathology continued to raise a considerable income by conducting routine investigations for local and county authorities and practitioners. Nationally, the troubled state of pathology was confirmed in 1924, when the Medical Research Council publicly derided pathologists for the relatively low quantity and quality of research. Nevertheless, many remained enveloped in the drudgery and commercialism of 'service' laboratories.[91] During the year ending 30 June 1931, specimens received by the department's lab came from thirty-four counties, county boroughs and district councils, forty-two hospitals and 470 practitioners, total reports exceeding 17,000. Examinations included 3,182 for venereal disease (including 2,276 Wassermann tests), 8,885 diphtheria swabs, 796 histological tests, as well as analyses of urine (275), pus

[88] BCLLS, Eye Hospital, Annual Report, 1937.
[89] BCLA, General Hospital, Medical Committee Minute Book, 1925–34, GHB 80.
[90] Ibid.
[91] Valier, 'The Politics of Scientific Medicine', 59–60.

(86), faeces (148), blood (552), milk (1,386, largely for TB), sputum (1,234), water (124), hair (70) and a bottle of beer. More importantly, the department's income approached £6,000, while costing the university only £3,585 (£2,473 of which comprised salaries).[92] By 1938, the department was earning the university £6,000 annually, only now members of the faculty agreed that its four academic staff and twelve technical assistants should spend less time conducting routine investigations, and undertake more research. In order to improve the department's research potential, the Department of Cancer Research was placed under the direction of the Professor of Pathology when its several labs, balance room and animal house were transferred to the new medical school in Edgbaston that same year.[93]

REFORM AND RELOCATE:
RECONFIGURING MEDICAL EDUCATION IN BIRMINGHAM

In general, although many changes were overdue and recognised as such by staff, it was the prospect of a new medical school that encouraged the most significant reorganisation to both teaching and research practices. Though an amalgamation of the city's two main hospitals was suggested as early as 1926, such proposals were resisted until the scheme appeared more certain. Two years later, however, the medical faculty was already discussing the new teaching centre in considerable detail, as its members' plans for the medical school indicate. Not surprisingly, the most space in the new building was set aside for anatomy, physiology and pathology. Anatomy alone was to have thirty-two rooms, and 20,750 square feet of floor space, while physiology occupied fifty rooms (22,500). Pathology was to have as many rooms as anatomy, though nine were to be located in the adjoining hospital. Therefore, of its 16,180 feet of ground space, 4,615 were in the hospital. Public health though appearing large (6,600) was to have a museum occupying 5,000 square feet and five additional offices occupied the remaining space. Pharmacology and forensic medicine, on the other hand, made do with the least space, being designated seven (4,350) and five (2,400) rooms respectively. General requirements, including the school library and periodical room, were allocated twenty rooms and 9,450 square feet. Tutorial rooms for students (3:960) and public health labs (17:5,020) made for a grand total of 171 rooms and 88,210 square feet.

By 1930, only minor revisions had been introduced to the plans. The total area of the new medical school had declined slightly to 86,617 square feet. General requirements comprised nearly 10,000 square feet, a mere 3,850 square feet were devoted to a library. The majority of space was still allotted

[92] UBSC, University of Birmingham, Faculty of Medicine Minutes, 1929–32.

[93] UBSC, University of Birmingham, Faculty of Medicine Minutes, 1936–8.

to anatomy and physiology, with 20,625 and 21,580 square feet respectively. Of the anatomy department, 4,000 square feet alone were for its museum, 2,500 square feet for a dissecting room, 1,200 square feet for a study room, and an x-ray apparatus was shared with physiology. Its half a dozen members of staff were also allocated a physical anthropological lab (450 square feet) and an advanced lab for embryology (450 square feet). With fourteen projected members of staff, pathology and bacteriology occupied considerably less space (16,700 square feet), though more than the lucrative university public health lab (5,022 square feet), pharmacology (4,350 square feet) and the department of public health 6,600 square feet (of which 5,000 square feet was for a museum alone). Forensic medicine, though also requiring space for lab work, occupied little more than 900 square feet. Medicine, surgery and midwifery, without museums and laboratories at the school, were each designated 960 square feet, but relied far more on clinical space in the teaching hospitals.[94]

By 1929, it was decided that clinical teaching in the new 600-bed Hospital Centre would include medicine, surgery, therapeutics, midwifery, diseases of women, ophthalmology, disease of ENT, orthopaedics, dermatology, venereal disease and radiology. Though a central hospital reduced the time students travelled between institutions, there were doubts about establishing every department in the new hospital, and the concept of associated hospitals continued to receive support. As a result, the Children's and Dental hospitals were declared departments in their own right. A Department of 'Mental Medicine' was also to be established with access to an outpatient clinic and 'beds for border-line Mental Cases' at the new hospital.[95]

For some academics, the long-awaited move to the Edgbaston campus coincided with a changed employment status. In general, the growth of the medical school, not to mention the increasing importance of research in these years, necessitated transforming ever more academic posts into full-time appointments. Though drastic, the clinical side of medical education in Birmingham had changed little over many decades. The appointment of whole-time assistants to improve clinical instruction, when introduced in 1933, for example, was described by members of faculty as the only fundamental change to clinical instruction in sixty years.[96] In 1936, more 'intimate work' required the appointment of additional staff.[97] Five years later, a Director of Clinical Research with control of thirty beds was also

94 UBSC, University of Birmingham, Faculty of Medicine Minutes, 1929–32.

95 UBSC, University of Birmingham, Faculty of Medicine Minutes, 1925–9.

96 UBSC, University of Birmingham, Faculty of Medicine Minutes, 1933–6.

97 UBSC, University of Birmingham, Faculty of Medicine Minutes, 1936–8.

appointed. In 1937 the first full-time chair of bacteriology was appointed. Previously, most staff had been paid an honorarium, which since the Great War had risen to £100. By the 1930s, specialisation had advanced to undreamed levels. Moreover, the influx of patients to hospital increased their unpaid work, while growth of the medical school similarly increased 'non-lucrative work'. By 1938, most part-time professors were spending half their time training students. Some medical schools solved this problem by appointing whole-time professors, with stipends ranging between £1,500 and £2,000. As each professor also required an assistant, this promised to cost the average university £7,000 annually. As a result, some schools had appointed their professors on a half-time basis and merely increased their stipend. Now that professors were required to organise departments and not just give lectures, stipends were increased to £250 or £500, though only £100 for forensic medicine as it occupied 'less and less of the curriculum'.[98]

The administration of the school was also becoming a full-time job for many members of staff. Appointed in 1931, Stanley Barnes was the first dean who was not the head of a department or a professor, as the faculty wished to appoint someone with more time to devote to the design and construction of the new medical school.[99] The demands of the post also required a clinician, as the dean organised clinical teaching and was not simply head of faculty. A pre-clinical man was also less likely to establish contact with 'men of light and leading' in Birmingham. Finally, the dean was required to attend civic functions and promote the school. Although assisted by three clerks by this time,[100] Barnes retired as honorary physician to the Birmingham United Hospital in 1936 owing to his increasing duties.[101] For his services, he received an annual salary of £500, as opposed to £100 previously. Though Barnes's name was not inscribed on any of the stones laid on the new medical building's site, by the time he retired in 1941, it was said his name was on every brick. According to the members of the medical faculty, the fact that he was dean during the medical school's move to Edgbaston made him the 'most important dean' to date.[102]

The medical school's transfer to the new Edgbaston site was a long and drawn out affair. Tickets for the opening of the new medical school were sent to 300 students and 900 alumni in May 1934. By October of that year, the foundation stone of the new hospital was laid by the Prince of Wales.[103]

[98] Ibid.

[99] Barnes, *Birmingham Hospitals Centre*, 73.

[100] UBSC, University of Birmingham, Faculty of Medicine Minutes, 1936–8.

[101] *Birmingham Post*, 10 June 1936.

[102] UBSC, University of Birmingham, Faculty of Medicine Minutes, 1938–41.

[103] *Birmingham Post*, 24 October 1934.

On 1 January 1935, the two original teaching hospitals were amalgamated by Act of Parliament and renamed the Birmingham United Hospital, but the medical school would take another three years to complete. By June 1936, however, the city's inhabitants were able to view a scale model which was exhibited in the local Council Room. Nearly two years later, the East and West blocks of the new school were advanced enough to permit the transfer of some departments. The new buildings were finally revealed to public inspection in the afternoon of 14 July 1938, when the medical school was officially opened by the King.

By 1941, only the central block of the school, with its library and main lecture theatre, were still to be completed. It was, however, already regarded as specially suited to research, being located in a 'city of a thousand trades'.[104] The school could cope with eighty new students a year, but Barnes suggested a reduction in their numbers be introduced after the war, as the school and hospital were designed for only sixty new admissions annually. Instead of concentrating on numbers, staff now selected applicants more carefully. In future, preference was to be given to those taking full degrees and coming from the Midlands.[105] What had started as a Midlands institution, very much remained one.

Despite a strong case for continuity, medical education in Birmingham changed considerably during the interwar period. In the years following the First World War, far greater efforts were made to introduce certain specialist subjects into clinical teaching. Inevitably, some specialties found their place in the undergraduate curriculum more quickly than others. For example, psychiatric medicine, venereal disease and maternal mortality all became priorities due to the proliferation of such cases during the war years. Meanwhile, the inclusion of many other subjects in the school's prospectus, especially paediatrics, relied on the initiative of influential practitioners, such as Leonard Parsons, before they achieved greater prominence. Many other subjects were rarely ever addressed by staff, except perhaps in a brief postgraduate course or lecture. Many others, however, became the subjects of innovative research projects, which improved existing ties between the medical school and its associated teaching hospitals.

Nevertheless, the way in which the medical school went about arranging clinical instruction demonstrates the calculated manner in which hospitals attained associate status. Despite the proliferation of hospitals in the first half of the twentieth century, many, particularly those run by local authorities, were not deemed worthy of associate status. As a result, medical school staff continued to face a shortage of teaching cases throughout this

[104] UBSC, University of Birmingham, Faculty of Medicine Minutes, 1938–41.
[105] Ibid.

period. If not immediately apparent, this was periodically reflected in students' examination results. Rather than improve links with existing institutions, including those in the municipal sector, medical school staff decided upon a different solution. To many, the answer to their immediate teaching difficulties lay in a new Hospital Centre in Edgbaston. Though not without its advantages, this particular solution only further distinguished the medical school and its associated hospitals from the region's other medical institutions. Neither did it manage to eliminate unnecessary duplication of medical services and the costs this entailed.

Conclusion

PRIOR TO 1779, very few of Birmingham's inhabitants entered hospitals in order to receive medical treatment. Indeed, hardly any of those living in the town had ever laid eyes on such a building. By the close of the 1930s, however, the city supported approximately two-dozen such medical institutions. Nationally, the number of voluntary hospitals approached 700.[1] Combined, these institutions formed a net of medical philanthropy which conceivably caught 'virtually everyone at one time or another'.[2] From 1779, when the General Hospital first filled its forty beds, medical institutions, including the second medical school in provincial England, multiplied and expanded steadily. By 1939, the nine earliest voluntary hospitals had become affiliated teaching hospitals and offered the town's sick and infirm more than 1,700 beds. These were occupied by nearly 30,000 inpatients during that year alone. Unlike the late eighteenth century, when the General Hospital treated very few outpatients, these nine voluntary hospitals now offered medical care to more than 130,000 patients. With voluntary hospital provision what it was in 1780, it would have taken medical staff at Birmingham's only hospital sixty years to treat a number of individuals equivalent to the town's population. By 1939, Birmingham's first nine voluntary hospitals could theoretically have treated the city's million inhabitants in just seven and a half years. Significantly, these calculations do not include the town's largest medical institution, the Dudley Road Hospital, the former workhouse infirmary. When the Birmingham Regional Health Board was established on 10 December 1946, its nine committees managed 220 hospitals and 42,000 beds.[3] With ever more women working, so the likelihood of patients remaining at home during illness also declined. Clearly, hospitals had become very familiar institutions to the inhabitants of Birmingham and its surrounding districts in less than two centuries. In turn, as ever greater numbers passed through Birmingham's teaching hospitals, medical practitioners and students became far more familiar with the afflictions of the wider community they served.

The institutions themselves changed significantly in the century following their first appearance. Resembling a large mansion when it first opened

[1] Pinker, *English Hospital Statistics*, 57.
[2] Prochaska, 'Philanthropy', 360.
[3] [Birmingham Regional Hospital Board], *Birmingham Regional Hospital Board*, 5.

in 1779, the General Hospital, equally domestic in its organisation, was more often referred to as a 'house', than a hospital. Subsequent institutions appeared similarly residential, though often took much longer to organise inpatient accommodation, most operating as dispensaries for years, or even decades, as was the case at the Dental Hospital. Once enlarged and equipped with furnished wards, these institutions generally offered the benefits of a good home, providing their inpatients with regular meals and considerable bed rest. Innovations when introduced to buildings were often very non-medical, though no less revolutionary, comprising indoor water closets, running water, baths, as well as gas and electric lighting. More often than not, hospitals were some of the first local institutions to adopt such modern conveniences.

More commonly addressed by historians, the type of health care these institutions offered their patients also changed significantly in this period. Noticeable in just the half century since the Skin Hospital's foundation (1881), medicine transformed at a very rapid pace. In the decade following the charity's establishment, cure was seen to depend on bacteriology, not bathing. While very few medical students had heard of bacteriology in 1880, as early as 1890 it was recognised as the most promising weapon against disease, the causative organisms of a host of illnesses, including TB and cholera, having been discovered in this decade alone. As a result of new laboratory techniques, many chronic cases and suspicious rashes were finally correctly diagnosed and treated; in the case of syphilis, a single dose of medication could turn a positive Wassermann reaction into a negative one, and, just as importantly in the eyes of hospital subscribers, permit a patient to resume employment.[4] The mass-production of other bacteriostatic agents during the 1940s would go on to inspire greater confidence in medicine among a post-Second World War public.

Though innovations were never incorporated as quickly as staff and governors often wished, change was apparent on the wards of the city's hospitals and the corridors of the medical school in the first years of the twentieth century. Originally conducting hundreds of tests in newly established pathology laboratories in the early twentieth century, by 1939, full-time pathologists, bacteriologists and biochemists at the Birmingham Hospitals Centre alone undertook more than 20,000 tests at a specially built pathology block. At four other voluntary hospitals, laboratory diagnoses had similarly transformed hospital medicine by 1939, medical staff at the Children's, Eye, Skin and Women's hospitals together performing more than 26,000

---

4 E. Assinder, 'Syphilis in the poorer classes: its diagnosis by the Wassermann Test and its incidence as demonstrated thereby, Part II', *Birmingham Medical Review* 76 (1914), 152.

tests. The workload of the average laboratory in subsequent decades would double every five years.[5] Though costing the specialist hospitals considerable sums, the performance of such tests helped the medical school fund further innovative methods, machines and manpower. Laboratory tests also helped staff more effectively control hospital infection by routinely screening all those patients who were admitted into local institutions.

Technological innovations that made the hospital more comfortable for patients also transformed medical care in this period, with x-rays being used more regularly to diagnose fractures, previously determined manually, as well as treat a variety of skin conditions and even cancers. In 1900, staff at all nine of Birmingham's first teaching hospitals combined took less than 1,000 x-rays. With so few cases to observe, radiological instruction took more than two decades to develop. By 1939, however, the number surpassed 26,000 exposures, allowing many practitioners and pupils to ponder patients' problems long after they had left a clinic or institution. Although technologies, such as the x-ray, and techniques, as practised in the laboratory, tended to distance practitioners from their patients,[6] the work of administrative staff and almoners reduced this tendency towards detachment by providing greater insight into patients' lives. Actual contact, on the other hand, also increased, with students attending ward rounds at most voluntary hospitals, and not just the two main teaching hospitals, in the 1930s. Treatment at the Birmingham teaching hospitals also literally involved more physical contact in these years, with 17,000 patients at the Children's, Orthopaedic and United Birmingham hospitals regularly attending for massage treatment.

Among its other consequences, the use of massage therapy further increased roles for women at hospitals during this period. Despite an increase in all staff appointments, the proportion of nurses to patients increased far more quickly than that of consultants. As importantly, far more women actually joined the honorary medical staffs at hospitals, and not only at paediatric and obstetric institutions, which traditionally attracted far greater numbers of female practitioners. By 1900, Birmingham began training a new generation of women to take over from this small group of pioneers. Female dispensers and anaesthetists were appointed even more widely. Nevertheless, by the conclusion of the 1930s, in addition to those women appointed to the Women's and Children's hospitals, female practitioners held posts at the Dental, Eye, Ear and Skin hospitals. Another half a dozen women filled various teaching appointments in

---

5 [Birmingham Regional Hospital Board], *Birmingham Regional Hospital Board*, 82.

6 S. J. Reiser, *Medicine and the Reign of Technology* (Cambridge, 1978), 230.

the departments of Physiology, Pathology, Bacteriology, Midwifery and Industrial Hygiene at the medical school.[7] Though resistance never disappeared entirely, greater representation on hospital management committees additionally helped modify masculine culture at all levels of the hospital service. No longer confined to the Samaritan Committee, women by 1939 gained representation on every one of the Hospital Centre's dozen committees, save that of the honorary chaplains. Unfortunately, the sole female members of the medical staff of Birmingham's premier teaching institution, besides the nurses, were three stipendiary anaesthetists and a single resident medical officer.[8] Further changes would have to wait until well after the Second World War.

The patient profile also changed significantly between 1779 and 1939. Originally drawing patients from neighbouring, as well as distant counties, hospitals in the interwar period were serving mainly Birmingham's inhabitants. This transformation in catchment areas was noticeable at general hospitals far earlier than at specialist institutions, residents from other Midlands towns having relied less and less on the General and Queen's hospitals throughout the nineteenth century, founding and supporting their own general hospitals instead. A similar pattern of hospital use has been observed in other regions.[9] Neighbouring towns, however, were more hesitant to support a full range of specialist institutions, not to mention a medical school. As a result, medical institutions that remained unique to the region continued to draw patients, as well as pupils, from the greatest distances.

Made more pleasant in the twentieth century and furnished with greater numbers of private wards, all hospitals did eventually begin to serve wider sections of the local community. Initially established to treat the 'deserving poor', voluntary hospitals had always attracted individuals outside this narrowly defined constituency. This was most apparent and usually tolerated at the specialist hospitals, especially the Birmingham Women's Hospital, where staff treated very serious illnesses and desperate patients from a diverse range of social backgrounds, many able to pay towards their treatment. According to annual reports and the capacity of new private wards, at least 2,000 private patients were gaining admission to these nine voluntary hospitals alone in the 1930s. Once discouraged from entrance, young and old were also proportionately represented in patient populations.

7 UBSC, University of Birmingham, Medicine Syllabus, 1938–9.

8 BCLLS, United Birmingham Hospital, Annual Report, 1939.

9 K. Webb, *From County Hospital to NHS Trust: The History and Archives of NHS Hospitals Services and Management in York, 1740–2000*, vol. 1: *History* (York, 2002), 149.

Besides advancing the work of certain specialists, war also considerably changed middle-class attitudes to hospitals.[10] Significantly, the dean of the medical school, Stanley Barnes, entered a private nursing home when recovering from illness in November 1931; two decades later he spent five weeks as an inpatient at the newly constructed Queen Elizabeth Hospital, albeit in a private room.[11] Many other practitioners sent their children to local hospitals even earlier.[12] Had the middle classes not been prepared to enter voluntary hospitals, few would have had access to numerous innovative services, such as radium therapy.[13] Machines similarly attracted a new class of patient to the hospital.[14] Once dependent on subscriptions, donations and the odd legacy, Birmingham's hospitals by the last decades of the nineteenth century raised a significant proportion of their income from patients' payments, and most famously from the Hospital Saturday Fund. As more and more working people invested in their own health care, hospitals themselves were finally able to invest in their own futures.

While medical patients comprised the majority of cases at voluntary hospitals at the beginning of the nineteenth century, surgical cases comprised ever greater proportions of patient populations over successive decades. Though local practitioners noted a decline in amputations, excisions of joints and lithotomies during the last quarter of the nineteenth century,[15] the range of surgery extended in most medical specialties in subsequent years. With an increase in surgical operations, periods of hospitalisation generally declined and turnover amongst inpatients also noticeably increased. In general, the average period of hospitalisation in Birmingham decline by about a third from the late nineteenth to the mid-twentieth centuries; a similar trend has been identified nationally.[16] By the end of the nineteenth century, surgical patients at general hospitals already outnumbered medical ones by a third, the greatest increases registering after the association between germs and infection was better understood.

Diagnoses, however, then as now, rose and fell like empires, depending on available technologies and individual staff appointments. As a result,

[10] Pickstone, *Medicine and Industrial Society*, 296.

[11] Barnes, *Birmingham Hospitals Centre*, 73, 116.

[12] E. T. Mathews, 'W. J. McCardie: the first specialist anaesthetist in the provinces', *Proceedings of the History of Anaesthetics Society* 11 (1992), 47.

[13] *Birmingham Post*, 15 January 1937.

[14] J. Howell, *Technology in the Hospital: Transforming Patient Care in the Twentieth Century* (Baltimore, 1995), 4–5.

[15] F. Jordan, 'On the gradual decrease in operative surgery', *Birmingham Medical Review* 5 (1876), 23–9.

[16] D. Armstrong, *A New History of Identity: A Sociology of Medical Knowledge* (London, 2002), 118.

medical personnel at the Ear Hospital might for a time have described themselves more appropriately as serving the local Nose Hospital. Similarly, had other charities considered renaming themselves based on the work they undertook, Birmingham in the early twentieth century might have been home to hospitals for tonsillectomy and circumcision. The case of the Skin Hospital, which never treated dermatological complaints alone, as well as the majority of provincial eye hospitals, which often treated great numbers of aural cases, equally demonstrates that the work undertaken at a hospital is not always clearly expressed in its name. In any case, ever increasing numbers of emergency patients would have added to the variety of conditions treated at all medical institutions over time. At the Eye Hospital, as well as the city's two main general hospitals, such casualties accounted for approximately half of all hospital admissions for much of the twentieth century, and their numbers only increased as accidents gradually replaced infection as the main killer of the region's young people.[17] At the first general meeting of the Birmingham Accident Hospital in 1942, it was estimated that 100,000 serious accidents in the city required hospital treatment annually.[18]

As the admission of ever greater numbers of emergency cases suggests, the story of Birmingham's hospitals and medical school became more local as the twentieth century progressed. While doctors and nurses may have been recruited from further afield at the end of our period than in the days of Ash and Cox, in the first decades of the twentieth century, patients more regularly came from the districts nearest the hospital, if not within the city boundary itself. The medical school, too, though proud of its international connections and reputation, recruited staff and students locally for much of this period. Its research similarly reflected local concerns throughout the twentieth century. Though noticeable in the fields of paediatrics and public health, it is perhaps best illustrated by the school's decision to found a department of Industrial Hygiene and Medicine in 1934.

While this was described as one of the region's most original contributions to medical research and local health care, other developments more clearly followed national trends. With the foundation of professional societies and improved communication between institutions nationally, this should not be surprising. Not only mirroring developments elsewhere, whether at other provincial schools, or leading specialist hospitals, the voluntary hospital system in Birmingham was characterised by much

---

[17] [Birmingham Regional Hospital Board], *Birmingham Regional Hospital Board*, 23.

[18] BCLA, Birmingham Accident Hospital, Board of Management Minutes, 1941–8, HC/AH/1/1/1.

duplication of services. Eager to raise their status, or at least provide a service worthy of a teaching hospital, the majority of hospitals aimed in the long
run to acquire their own x-ray departments and pathology labs. During the
interwar period, the replication of services would have proved even more
expensive, even wasteful, though financial difficulties at many institutions
regularly encouraged at least periods of co-operation. Radium, for example,
was shared between hospitals, and a joint massage school was supported
by the Orthopaedic and General hospitals. Occasionally, this went much
further, as when the Cripples Union and Orthopaedic hospitals merged,
or the Women's and Maternity charities amalgamated. Equally, much co-
operation was encouraged by the medical school, a more neutral focus for
reform, especially as its prestige grew in the interwar period. Despite the
impressive results of numerous collective efforts, governors at most local
medical institutions charted their courses independently and guarded their
legacies jealously. As a result, many services and facilities continued to exist
in duplicate, triplicate, and so forth.

Where duplication was most widespread, not surprisingly, administrators faced the greatest financial difficulties at the end of our period. The
work of the Ear Hospital, for example, was initially very similar to that
carried out at the Eye Hospital and even for a time replicated at another
local Ear Hospital. In the twentieth century, with the formation of specialist clinics at the Children's, General and Queen's hospitals, which took over
the aural charity's teaching function, its mission became even less clear.
The same was true of the Skin Hospital. Much of its venereal work was
duplicated at the Queen's Hospital, but similar clinics also soon appeared
at the Children's and Women's hospitals. While dermatological work
took up more of the staff's work in the twentieth century, skin departments were established at the two local general hospitals, as well as the
Children's Hospital. Equally, though the Orthopaedic Hospital might have
been regarded as similarly wasteful in its early years, governors were more
successful in defining a clear role for the charity in the twentieth century.
Nevertheless, the majority of these institutions survived well after their
incorporation into the National Health Service. Despite taking some direct
hits in the Second World War, three in the case of the Skin Hospital,[19]
closure, for many, came only in the 1990s.

As in more recent years, co-ordination never went as far as some might
have expected, or wished. The Joint Hospital Scheme and Hospitals
Council discouraged some building schemes and encouraged the better
organisation of appeals. Equally, the return of war in 1939 favoured the
multiplication of additional co-operative efforts. The contributory scheme,

---

[19] UBSC, Skin Hospital, Annual Reports, 1940, 1942.

the new Hospital Centre in Edgbaston and the Emergency Medical Service only took co-ordination to new levels. However, the amalgamation of two general hospitals in the 1930s simply led another to spring up where the Queen's Hospital had once existed. To have expected otherwise is to forget the context in which these institutions emerged. The voluntary system, after all, was predicated on individual initiative and little government involvement. In fact, the majority of the hospitals covered in this volume, whether specialist or general, were, like the medical school, the result of a single person's initiative. In subsequent years, their development was equally determined by a small group of committed individuals, often spending funds donated by only a few hundred local families. Whether judged by the appearance of the buildings, the medical services on offer, or the ailments of patients, Birmingham's teaching hospitals in every respect were products of their time and place. Just as this context changed in 1948, so, too, did the overriding principles at the heart of the system transform in the middle of the twentieth century. Still part of the city's landscape today, many of these buildings continue to remind us of the needs and achievements of a previous generation. Similarly, their disappearance, if not their transformation into non-medical institutions, says as much about those of a post-industrial Birmingham.

Hospital locations

General Hospital
A1 1779-1896 Summer Lane
A2 1897- Steelhouse Lane

Orthopaedic Hospital
B1 1817- New Street
B2 1824-57 89 New Street, home of Martin Nobel Shipton, surgeon to the hospital
B3 1857-77 21 Great Charles Street
B4 1877-98 81 Newhall Street (corner of Great Charles Street)
B3 1898-1927 22 Great Charles Street
B5 *c.* 1900 convalescent home Islington Row, Edgbaston
    1909- Woodlands, Bristol Road (added to hospital in 1925)
    1920 Forelands in Bromsgrove
    1925 amalgamation with Cripples Union and acquire
B6 1925- Broad Street (old Children's Hospital purchased)
    1938 Vicarage Road site closed (patients moved to Woodlands)
    by 1930s Woodlands, Forelands and Broad Street remaining sites.

Eye Hospital
C1 1823- 35 Cannon Street (rebuilt 1838)
C2 1849- Steelhouse Lane (becomes the Children's Hospital in 1861)
C3 1861-83 Dee's Royal Hotel (Ann Street)
C4 1883- corner of Edmund Street and Church Street

Medical School (founded by Sands Cox)
D1 1825-8 Temple Row
D2 1828-36 Snow Hill Station
D3 1836-1938 Paradise Street
    1892 merged with Mason's College
D4 1900-38 occupied more rooms in Mason's College when most of college moved to Edgbaston
    1938- Vincent Drive, Edgbaston

Medical School (Sydenham College – rival school)
E1 1851-8 St Paul's Square
E2 1858-68 Steelhouse Lane (opposite General Hospital)
D3 1868 schools merged and moved the Paradise Street building

Queen's Hospital
F 1841- Bath Row

Ear Hospital (not to be confused with second ear hospital based at Cambridge and Suffolk streets during this period)
G1 1844-68? Cherry Street
G2 1868-77 Ann Street
G3 1877-83 81 Newhall Street (with Orthopaedic Hospital)
G4 1884-91 7 Great Charles Street
G5 1891 Edmund Street

Dental Hospital
H1 1858 13 Temple Street
H2 1863- 2 Upper Priory (shared with Homeopathic Hospital)
H3 1871-82 9 Broad Street
H4 1882-1906 71 Newhall Street
H5 1906- Great Charles Street

Children's Hospital
I1 1861-70 Steelhouse Lane
I2 1870-1916 Broad Street (outpatient department remained in Steelhouse Lane)
I3 1916- Ladywood Road

Skin Hospital
J1 1881-6 Newhall Street
J2 1886- John Bright Street (remains outpatient department after 1935)
J3 1935- George Street, Edgbaston (new inpatient hospital)

Women's Hospital
K1 1871-78 8 The Crescent
    1878-1905 Sparkhill
    1905 Showell Green Lane

Maternity Hospital
L1 1910 Loveday Street

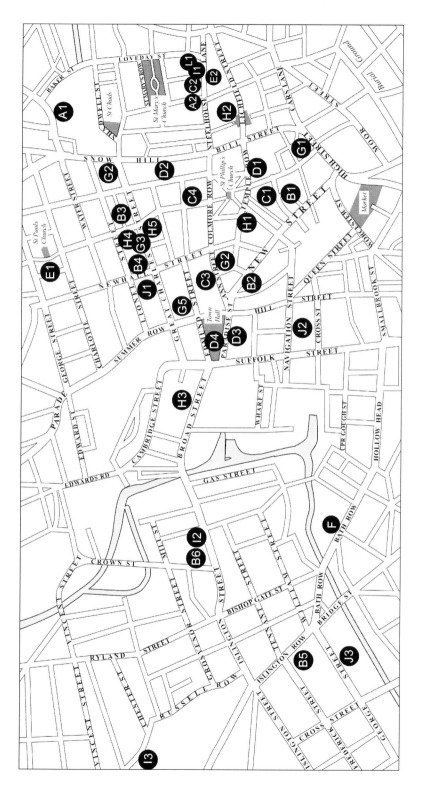

Patient Numbers at the Hospitals, 1780–1939

INPATIENTS

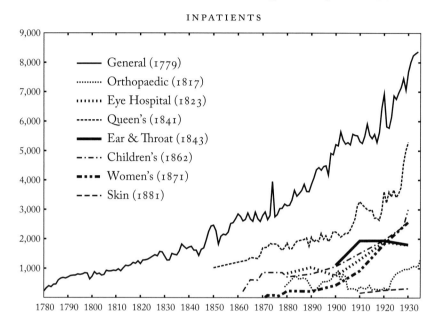

Legend:
— General (1779)
.... Orthopaedic (1817)
▪▪▪▪▪▪▪ Eye Hospital (1823)
------- Queen's (1841)
━━━ Ear & Throat (1843)
-·-·- Children's (1862)
▬▬▬ Women's (1871)
- - - Skin (1881)

OUTPATIENTS

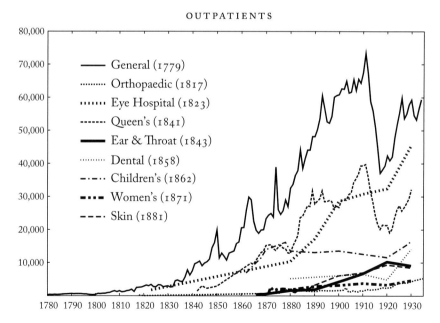

Legend:
— General (1779)
.... Orthopaedic (1817)
▪▪▪▪▪▪▪ Eye Hospital (1823)
------- Queen's (1841)
━━━ Ear & Throat (1843)
.... Dental (1858)
-·-·- Children's (1862)
▬▬▬ Women's (1871)
- - - Skin (1881)

Sources: BCLA, Annual Reports: Children's Hospital, 1862–1930; Dental Hospital (only outpatients treated in this period), 1880–1930; Ear & Throat Hospital, 1866–1930; Eye Hospital, 1823–1930; General Hospital, 1780–1934; Orthopaedic Hospital, 1818–1939; Queen's Hospital, 1842–1930; Skin Hospital, 1881–1930; Women's Hospital, 1880–1930

Bibliography

Unpublished Primary Sources

BIRMINGHAM CITY LIBRARY ARCHIVES (BCLA)
Records of Birmingham Accident Hospital, HC/AH (MS 1793)
Records of Birmingham Board of Guardians, GP/B
Records of Birmingham Children's Hospital, HC/BCH (MS 1439)
Records of Birmingham Dental Hospital, HC/DH/2006/105
Records of Birmingham and Midland Eye Hospital, MS 1919 (roughly sorted)
Records of Birmingham and Midland Hospital for Women, HC/WH (MS 1940)
Records of Birmingham and Midland Skin Hospital, MS 1918 (uncatalogued)
Records of the Birmingham Lying-In Hospital, MH
Records of the General Hospital, Birmingham, HC/GH (MS 1921, cataloguing in process)
Records of the Queen Elizabeth Hospital, HC/QE (MS 1942)
Records of the Queen's Hospital, HC/QE (MS 1942)
Records of the Royal Orthopaedic Hospital, HC/RO

BIRMINGHAM CITY LIBRARY, LOCAL STUDIES SECTION (BCLLS)
Annual Reports of the Birmingham and Midland Eye Hospital
Annual Reports of the Birmingham and Midland Hospital for Women
Annual Reports of the Birmingham and Midland Skin Hospital
Annual Reports of the Birmingham Ear Hospital
Annual Reports of the General Hospital, Birmingham
Annual Reports of the Queen's Hospital, Birmingham
Annual Reports of the Royal Orthopaedic Hospital
Newspaper cuttings relating to Birmingham hospitals (2 volumes), 1863–1914.

UNIVERSITY OF BIRMINGHAM SPECIAL COLLECTIONS (UBSC)
Minutes of Sydenham College, Birmingham, 1851–66, MS 128/1
Records of the Birmingham Dental Hospital
Records of the University of Birmingham Medical School
Vaughan Thomas Collection, MSS 281/i/10

Published Primary Sources

Select Committee on Medical Education, 1834, PP xiii

NEWSPAPERS AND PERIODICALS
Aris's Gazette
Birmingham Journal
Birmingham Mail
Birmingham Post

Birmingham Red Book and Reference Almanac
British Journal of Dental Science
British Medical Journal
Dental Record
Hospital
Johannesburg Star
Lancet

CONTEMPORARY BOOKS AND ARTICLES

Alanson, E., *Practical Observations on Amputation*. London, 1782

[Anon.], *History of the Wolverhampton and Midland Counties Eye Infirmary, 1881–1931*. Wolverhampton, 1931

Ash, J., *Experiments and Observations, to Investigate, by Chemical Analysis, the Medicinal Properties of the Mineral Waters of Spa and Aix-la-Chappelle, in Germany; and of the Waters and Boue near St. Amand in French Flanders*. London, 1788

Assinder, E., 'Syphilis in the Poorer Classes: its diagnosis by the Wassermann Test and its incidence as demonstrated thereby, Part II', *Birmingham Medical Review* 76 (1914), 137–53

Bateman, T., *A Practical Synopsis of Cutaneous Diseases, According to the Arrangement of Dr Willan*. London, 1836

Bell, B., *A Treatise on the Theory and Management of Ulcers*. Edinburgh, 1777

[Birmingham Cripples Union and Royal Orthopaedic and Spinal Hospital], *Birmingham Cripples Union and Royal Orthopaedic and Spinal Hospital, 1817–1915*. Birmingham, 1925

Budd, W., 'On intestinal fever', *Lancet* 2 (1859), 4–5, 28–30, 55–6, 80–2

Bulleid, W. A., 'The separation of dentistry from medicine', *British Dental Journal* 62 (1937), 113–21

Bunce, J. T., *Birmingham General Hospital, 1779–1897*. Birmingham, 1897

Cohen, R., 'English Theories of the causes of dental caries, 1800–1850', *Dental Magazine and Oral Topics* 53, 6 and 7 (1936), 1–15

Darwall, J., *Plain Instructions for the Management of Infants, with Practical Observations on the Disorders Incident to Children*. London, 1830

Dufton, W., *The Nature and Treatment of Deafness and Diseases of the Ear*. Birmingham, 1844

Eales, H., 'The treatment of cataract', *Birmingham Medical Review* 27 (1890), 257–67

Eliot, G., *Middlemarch*. London, 1994

Galton, F., *Memories of My Life*. London, 1908

Gray, J., *Dental Practice*. London, 1837

Griffiths, G., *History of the Free-Schools, Colleges, Hospitals, and Asylums of Birmingham*. London, 1861

Gutteridge, T., *The Crisis: Another Warning Addressed to the Governors* [General Hospital, Birmingham]. Birmingham, 1851

Harrison, E., *Pathological and Practical Observations on Spinal Diseases*. London, 1827

Heslop, T. P., *The Realities of Medical Attendance on the Sick Children of the Poor in Large Towns*. London, 1869

Hodgson, J., *A Treatise on the Diseases of Arteries and Veins*. London, 1815

Howard, J., *An Account of the Principal Lazarettos in Europe*. London, 1789

Hume, G. H., *The History of the Newcastle Infirmary*. Newcastle-upon-Tyne, 1906

Hutton, W., *History of Birmingham*. Birmingham, 1781

[The Inquirer], 'On the treatment of ulcerated legs', *Edinburgh Medical and Surgical Journal* 1 (April 1805), 187–93

Jones, J. E., *A History of the Hospitals and other Charities of Birmingham*. Birmingham, 1908

Jordan, F., 'On the gradual decrease in operative surgery', *Birmingham Medical Review* 5 (1876), 23–9

Kershaw, R., 'British ear and throat clinics historically considered', *Journal of Laryngology, Rhinology and Otology* (1913), 421–7

Leader, J. D., *Sheffield General Infirmary*. Sheffield, 1897

Lindsay, L., *A Short History of Dentistry*. London, 1933

Mackey, E., 'Remarks upon the special study of skin diseases; with an analysis of three hundred cases', *Birmingham Medical Review* 4 (1875), 240–5

McKay, W. J. S., *Lawson Tait: his Life and Work*. London, 1922

Middlemore, R., *A Treatise on the Diseases of the Eye and its appendages*. 2 vols. London, 1835

Moore, J., *Clinical Reports of Surgical Cases under the treatment of W. Sands Cox, 1843–4*. Birmingham, 1844

Morris, M., 'An address delivered at the opening of the Section of Dermatology', *British Medical Journal* 2 (18 Sept. 1897), 697–702

Morrison, J. T. J., *William Sands Cox and the Birmingham Medical School*. Birmingham, 1926

Newman, G., *Recent Advances in Medical Education in England: a memorandum addressed to the Minister of Health*. London, 1923

Parker, L., *The Modern Treatment of Syphilitic Disease*. London, 1859

Parker, S. A., *Remarks upon Artificial Teeth*. Birmingham, 1862

— 'Contributions to dental surgery', *British Journal of Dental Science*, 6, 84 (1863), 263–8

Parks, N., 'Dentistry of fifty years ago', *The Dental Record*, 22, 11 (1902), 487–98

Percival, T., *Medical Ethics*. Manchester, 1803

Pott, P., *Some Few General Remarks on Fractures and Dislocations*. London, 1796

Pye, C., *A Description of Modern Birmingham*. Birmingham, 1819

Richardson, B. W., 'Hygeia, a city of health', *Nature* (12 October 1876), 542–5

Robertson, W., *Practical observations on the teeth*. London, 1846

Robinson, J., *Dental Anatomy and Surgery*. London, 1846

Rowley, W., *A Treatise on the Cure of Ulcerated Legs without Rest*. London, 1774

Ryland, F., *A Treatise on the Diseases and Injuries of the Larynx and Trachea*. London, 1837

Savage, T., 'On removal of the uterine appendages', *British Medical Journal* (8 Jan 1887), 51–2

[Scutator], *The Medical Charities of Birmingham: being letters on hospital adminstration and management*. Birmingham, 1863

Senex, 'A reminiscence of dentistry in the old days: a fortunate dental student, 1855', *British Journal of Dental Science* 68, 1225 (1920), 6–12

Simpson, J. Y., 'Report of the Edinburgh Royal Maternity Hospital, St John's Street', *Monthly Journal of Medical Science* 9 (1848–9), 329–38

Smith, A., *The Wealth of Nations* (1776). London, 1987

Smith, G. M., *A History of Bristol Royal Infirmary*. London, 1917

Smith, P., *On the Pathology and Treatment of Glaucoma*. London, 1891

Syme, J., *Treatise on the Excision of Diseased Joints*. London, 1831

Tait, L., *Diseases of Women*. New York, 1879

Tait, R. L., *Diseases of Women and Abdominal Surgery*. Leicester, 1889

Taylor, M., *Sir Bertram Windle: a Memoir*. London, 1932

Thomas, V., *An address upon laying the foundation-stone of the Queen's Hospital, Birmingham, June 18, 1840*. Oxford, 1840

Tomlinson, T., *Medical Miscellany*. London, 1769, republished 1774

Turner, D., *A Treatise of Diseases Incident to the Skin*. London, 1736

Turner G. G., and W. D. Arnison, *The Newcastle upon Tyne School of Medicine, 1834–1934*. Newcastle upon Tyne, 1934

Underwood, M., *Surgical Tracts, Including a Treatise upon Ulcers of the Legs*. London, 1788

Waddy, J. M., 'Report of the Birmingham Lying-In Hospital, and Dispensary for the Diseases of Women and Children', *Provincial Medical and Surgical Journal* 9, 3 (1845), 39–41

Webb, S., 'The economic theory of a legal minimum wage', *Journal of Political Economy* 20, 10 (1912), 973–98

Wellings, A. W., 'John Humphreys, of Birmingham: an appreciation', *British Dental Journal* 53, 1 (1937), 49–50

Windle, B. and Hillhouse, W., *The Birmingham School of Medicine*. Birmingham, 1890

Withering, W., *An Account of the Foxglove and Some of its Medical Uses: with Practical Remarks on Dropsy and Other Diseases*. London, 1785

Worth, C., *Squint, its Causes, Pathology and Treatment*. London, 1903

Young, F. B., *The Young Physician*. London, 1938

SECONDARY SOURCES

Abel-Smith, B., *A History of the Nursing Profession*. London, 1960

——— *The Hospitals, 1800–1948*. Cambridge, 1964

Ackerknecht, E., *Medicine at the Paris Hospital, 1794–1848*. Baltimore, 1967

Allan, R. N., and M. G. Fitzgerald, eds, *Historical Sketches in the West Midlands*, vol. 2. Shipston-on-Stour, West Midlands, 1984

Andrew, D., 'Two medical charities in eighteenth-century London: The Lock Hospital and the Lying-in Charity for Married Women', in *Medicine and Charity before the Welfare State*, ed. J. Barry and C. Jones. London, 1991

Andrews, C. T., *The First Cornish Hospital*. Penzance, 1975

Anning, S. T., 'Provincial Medical Schools in the Nineteenth Century', in *The Evolution of Medical Education in Britain*, ed. F. N. L. Poynter. London, 1966

Armstrong, D., *A New History of Identity: A Sociology of Medical Knowledge*. London, 2002

Austoker, J., and L. Bryder, eds, *Historical Perspectives on the Role of the MRC: essays in the history of the Medical Research Council of the United Kingdom and its predecessor, the Medical Research Committee, 1913–1953*. Oxford, 1989

Austin, R. T., *Robert Chessher of Hinckley, 1750–1831: The First English Orthopaedist*. Leicester, 1981

Barnes, S., *The Birmingham Hospitals Centre*. Birmingham, 1952

Barry, J., and C. Jones, C., *Medicine and Charity before the Welfare State*. London, 1991

Bell, E. M., *Storming the Citadel: The Rise of the Woman Doctor*. London, 1953

Blake, C., *The Charge of the Parasols: Women's Entry to the Medical Profession*. London, 1990

Bonnell, J. A., 'Donald Hunter', *Journal of the Society of Occupational Medicine* 29, 2 (1979), 81

Bonner, T. N., *To the Ends of the Earth: Women's Search for Education in Medicine*. Cambridge, MA, 1992

Booth, C. C., 'Half a century of science and technology at Hammersmith', *British Medical Journal* 291 (1985), 1771–9

Borsay, A., *Disability and Social Policy in Britain since 1750*. Basingstoke, 2005

Burrows, E. H., *Pioneers and Early Years: A History of British Radiology*. Alderney, 1986

Cantor, D., 'The MRC's support for experimental radiology during the inter-war years', in *Historical Perspectives on the Role of the MRC: essays in the history of the Medical Research Council of the United Kingdom and its predecessor, the Medical Research Committee, 1913–1953*, ed. J. Austoker and L. Bryder. Oxford, 1989

—— 'The contradictions of specialisation: rheumatism and the decline of the spa in inter-war Britain', in *The Medical History of Waters and Spas*, ed. R. Porter. London, 1990

Cavallo, S., 'Charity, power, and patronage in eighteenth-century Italian hospitals: the case of Turin', in *The Hospital in History*, ed. L. Granshaw and R. Porter. London, 1989

Cherry, S., 'The hospitals and population growth, Part 1: The voluntary general hospitals, mortality and local populations in the English provinces in the eighteenth and nineteenth centuries', *Population Studies* 34 (1980), 59–76

—— 'Beyond National Health Insurance: The voluntary hospitals and hospital contributory schemes: a regional study' *Social History of Medicine* 5, 3 (1992), 455–82

—— 'Before the National Health Service: financing the voluntary hospitals, 1900–1939', *Economic History Review* 50, 2 (1997), 305–26

—— 'Hospital Saturday, workplace collections and issues in late nineteenth-century hospital funding', *Medical History* 44, 4 (2000), 461–88

Cohen, R., 'The Birmingham Dental Hospital: with some account of the founding of the Dental School', *Birmingham Medical Review* 20 (1958), 331–7

—— Marsland, E. A. and Hillam, C., 'William Robertson of Birmingham 1794–1870', *British Dental Journal* 142 (1977), 64–9

Collier, H. E., *Experiment with a Life*. Wallingford, PA, 1953

Cooter, R., *Surgery and Society in Peace and War: Orthopaedics and the Organisation of Modern Medicine, 1880–1948*. Basingstoke, 1993

Cope, Z., *The Royal College of Surgeons of England*. London, 1959

—— 'The history of the dispensary movement', in *The Evolution of the Hospitals in Britain*, ed. F. N. L. Poynter. London, 1967

Corlett, H., '"No small uncertainty": eye treatments in eighteenth-century England and France', *Medical History* 42 (1998), 217–34

Davidson, L., '"Identities ascertained": British ophthalmology in the first half of the nineteenth century', *Social History of Medicine* 9, 3 (1996), 313–33

Dunn, P., 'Sir Leonard Parsons of Birmingham (1879–1950) and antenatal paediatrics', *Archives of Disease in Childhood* 86, 1 (2002), 65–7

Dyhouse, C., *No Distinction of Sex?: Women in British Universities, 1870–1939*. London, 1995

Fissell, M., *Patients, Power and the Poor in Eighteenth-Century Bristol*. Cambridge, 1991

Forbes, E. G., 'The professionalization of dentistry in the United Kingdom', *Medical History* 29 (1985), 169–81

Foster, W. D., *Pathology as a Profession in Great Britain and the Early History of the Royal College of Pathologists*. London, 1982

Frizelle, E. R., and J. D. Martin, *The Leicester Royal Infirmary, 1771–1971*. Leicester, 1971

Garrison F. H., and A. F. Abt, *Abt-Garrison History of Paediatrics, with New Chapters on the History of Paediatrics in Recent Times*. London, 1965

Goodman, G. M., *A Victorian Surgeon: A Biography of James Fitzjames Fraser West, 1833–83, Birmingham Surgeon*. Studley, 2007

Gorsky, M., J. Mohan and M. Powell, 'The financial health of voluntary hospitals in interwar Britain', *Economic History Review* 53, 3 (2002), 533–57

Gorsky, M., J. Mohan and T. Willis, 'Hospital Contributory Schemes and the NHS Debates 1937–46: The rejection of social insurance in the British Welfare State?' *Twentieth Century British History* 16, 2 (2005), 170–92

Gorsky, M., and J. Mohan, with T. Willis, *Mutualism and Health Care: British Hospital Contributory Schemes in the Twentieth Century*. Manchester, 2006

Gould, G., *A History of the Royal National Throat, Nose and Ear Hospital, 1874–1982*. Ashford, 1998

Granshaw, L., *St Mark's Hospital, London: A Social History of a Specialist Hospital*. London, 1985

—— and R. Porter, eds, *The Hospital in History*. London, 1989

Granshaw, L., 'Fame and fortune by means of bricks and mortar': the medical profession and specialist hospitals in Britain, 1800–1948', in *The Hospital in History*, ed. L. Granshaw and R. Porter. London, 1989

Hadley, R. M., 'The life and work of Sir William James Erasmus Wilson, 1809–84', *Medical History* 3 (1959), 215–47

Hardy, A., *Health and Medicine in Britain since 1860*. Basingstoke, 2001

Harris, B., *The Health of the Schoolchild: A History of the School Medical Service in England and Wales*. Buckingham, 1995

—— *The Origins of the British Welfare State: Social Welfare in England and Wales, 1800–1945*. Basingstoke, 2004

Heaman, E., *St Mary's: The History of a London Teaching Hospital*. London, 2003

Hearn, G. W., *Dudley Road Hospital, 1887–1987*. Birmingham, 1987

Hillam, C., *Brass Plate and Brazen Impudence: Dental Practice in the Provinces, 1755–1855*. Liverpool, 1991

Hodgkinson, R. G., *The Origins of the National Health Service*. London, 1967

Hopkins, E., *Birmingham: The First Manufacturing Town in the World, 1760–1840*. London, 1989

—— *Working-class Self-Help in Nineteenth-Century England: Responses to Industrialization*. London, 1995

—— *The Rise of the Manufacturing Town: Birmingham and the Industrial Revolution*. Stroud, 1998

—— *Birmingham: The Making of the Second City, 1850–1939*. Stroud, 2001

Howell, J., *Technology in the Hospital: Transforming Patient Care in the Twentieth Century*. Baltimore, 1995

Huisman, F. and Warner, J. H., 'Medical Histories', in *Locating Medical History: The Stories and Their Meanings*, ed. J. Huisman and J. H. Warner. Baltimore, 2004

Humphreys, R., *Sin, Organized Charity and the Poor Law in Victorian England*. Basingstoke, 1995

Ives, E., D. Drummond and L. Schwarz, *The First Civic University: Birmingham, 1880–1980: An Introductory History*. Birmingham, 2000

Jackson, M., 'Allergy and history', *Studies in History and Philosophy of Biological and Biomedical Sciences* 34, 3 (2003), 383–98

Jacob, F. H., *A History of the General Hospital near Nottingham*. Bristol, 1951

Jones, H., *Health and Society in Twentieth-Century Britain*. London, 1994

Jordan, E., '"Suitable and remunerative employment": the feminization of hospital dispensing in late-nineteenth-century England', *Social History of Medicine* 15, 3 (2002), 429–56

King, R., *The History of Dentistry: Technique and Demand*. Cambridge, 1997

Lane, J., *A Social History of Medicine*. London, 2001

Lansbury, C., *The Old Brown Dog: Women, Workers and Antivivisection in Edwardian England*. Madison, WI, 1985

Law, F. W., *The History and Traditions of Moorfields Eye Hospital*, vol. 2. London, 1975

Lawrence, C., 'A tale of two sciences: bedside and bench in twentieth-century Britain', *Medical History* 43 (1999), 421–49

LeVay, D., *The History of Orthopaedics*. Carnforth, 1990

Lewis, J., *The Politics of Motherhood: Child and Maternal Welfare in England, 1900–39*. London, 1980

Lockhart, J., 'Truly, a hospital for women': The Birmingham and Midland Hospital for Women', in *Medicine and Society in the Midlands, 1750–1950*, ed. J. Reinarz. Birmingham, 2007

Lomax, E. M. R., Small and special: the development of hospitals for children in Victorian Britain. London, 1996

Loudon, I. S. L., 'Leg ulcers in the eighteenth and early nineteenth centuries', *Journal of the Royal College of General Practitioners* 31 (1981), 263–73

—— *Death in Childbirth: An International Study of Maternal Care and Maternal Mortality, 1800–1950*. Oxford, 1982

—— 'On maternal and infant mortality, 1900–60', *Social History of Medicine* 4 (1991), 29–73

McMenemey, W. H., *A History of the Worcester Royal Infirmary*. London, 1947

—— 'William Sands Cox and the stoicism of Elizabeth Powis', *Medical History* 2 (1958), 109–13

Malins, J. M., 'A history of the Birmingham General Hospital', *Midland Medical Review* 11 (1975), 18–21

—— 'The General Hospital, Birmingham', in *Historical Sketches in the West Midlands*, vol. 2, ed. R. N. Allan and M. G. Fitzgerald. Shipston-on-Stour, West Midlands, 1984

Mamminga, M., 'British brass bands', *Music Educators Journal* 58, 3 (1971), 82–3

Manton, J., *Elizabeth Garrett Anderson*. London, 1965

Marland, H., *Medicine and Society in Wakefield and Huddersfield*. Cambridge, 1987

—— *Doncaster Dispensary, 1792–1867: Sickness, Charity and Society*. Doncaster, 1989

—— '"Pioneer work on all sides": the first generations of women physicians in the Netherlands, 1879–1930', *Journal of the History of Medicine and Allied Sciences* 50, 4 (1995), 441–77

Mathews, E. T., 'W. J. McCardie: the first specialist anaesthetist in the provinces', *Proceedings of the History of Anaesthetics Society* 11 (1992), 45–9

—— 'Startin's pneumatic inhaler', *Proceedings of the History of Anaesthesia Society* 32 (2003), 20–6

Maulitz, R. C., *Morbid Appearances: The Anatomy of Pathology in the Early Nineteenth Century*. Cambridge, 1987

Money, J., *Experience and Identity: Birmingham and the West Midlands, 1760–1800*. Manchester, 1977

Mooney, G., B. Luckin and A. Tanner, 'Patient pathways: solving the problem of institutional mortality in London during the later nineteenth century', *Social History of Medicine* 12, 2 (1999), 227–69

Mooney, G., and J. Reinarz, eds, *Permeable Walls: Historical Perspectives on Hospital and Asylum Visiting.* Amsterdam, 2009

Moscucci, O., *The Science of Woman: Gynaecology and Gender in England, 1800–1929.* Cambridge, 1990

Murray, J. J., Murray, I. D. and Hill, B., *Newcastle Dental School and Hospital, 1895–1995.* Newcastle, 1995

O'Connor, W. J., *Founders of British Physiology: A Biographical Dictionary, 1820–1885.* Manchester, 1988

O'Flynn, K., ' "Intellectual athletes": changing selection criteria for London medical schools from 1918–1939 to 2002', *Journal of History in Higher Education* 73 (2004), 11–23

O'Malley, C. D., ed., *The History of Medical Education.* Los Angeles, 1970

Peterson, M. J., *The Medical Profession in Mid-Victorian London.* Berkeley, CA, 1978

Pickstone, J. V., *Medicine and Industrial Society: a History of Hospital Development in Manchester and its Region.* Manchester, 1985

Pinker, R., *English Hospital Statistics, 1861–1938.* London, 1966

Pollard, S., *The Development of the British Economy, 1914–1950.* London, 1962

Porter, R., 'The gift relation: philanthropy and provincial hospitals in eighteenth-century England', in *The Hospital in History*, ed L. Granshaw and R. Porter. London, 1989

—— ed., *The Medical History of Waters and Spas.* London, 1990

—— *The Greatest Benefit to Mankind.* London, 1997

Poynter, F. N. L, ed., *The Evolution of the Hospitals in Britain.* London, 1967

—— 'Medical education in England since 1600', in *The History of Medical Education*, ed. C. D. O'Malley. Los Angeles, 1970

Prochaska, F. K., *Women and Philanthropy in Nineteenth-Century England.* Oxford, 1980

—— *Philanthropy and the Hospitals of London: The King's Fund, 1897–1990.* Oxford, 1992

—— 'Philanthropy', in *The Cambridge Social History of Britain, 1750–1950*, vol. 3, ed. F. M. L. Thompson. Cambridge, 1990

Reinarz, J., *The Birth of a Provincial Hospital: the Early Years of the General Hospital, Birmingham, 1765–1790.* Dugdale Society Occasional Paper 55. Stratford, 2003

—— 'The age of museum medicine: the rise and fall of the medical museum at Birmingham's Medical School', *Social History of Medicine* 18, 3 (2005), 419–37

—— and A. Williams. 'John Darwall, MD (1796–1833): the short yet productive life of a Birmingham practitioner', *Journal of Medical Biography* 13 (2005), 150–4

—— 'Investigating the "deserving" poor: Charity, discipline and voluntary hospitals in nineteenth-century Birmingham', in *Reconfiguring the Recipient: Historical Perspectives on the negotiation of medicine, charity and mutual aid*, ed. A. Borsay and P. Shapely. Aldershot, 2007

—— ed., *Medicine and Society in the Midlands, 1750–1950.* Birmingham, 2007

—— 'Unearthing and dissecting the records of English provincial medical education, *c.* 1825–1948', *Social History of Medicine* 21 (2008), 381–92

Richardson, R., *Death, Dissection, and the Destitute.* Chicago, 2001

Risse, G., *Hospital Life in Enlightenment Scotland.* Cambridge, 1986

Roberts, S., *Sophia Jex-Blake: A Woman Pioneer in Nineteenth Century Medical Reform.* London, 1993

Rook, A., 'James Startin, Jonathan Hutchinson and the Blackfriars Skin Hospital', *British Journal of Dermatology* 99 (1978), 215–19

—— 'Dermatology in Britain in the late nineteenth century', *British Journal of Dermatology* 100, 1 (1979), 3–12

—— Carleton, M., and W. G. Cannon, *The History of Addenbrooke's Hospital.* Cambridge, 1991

Rosen, G., 'Changing attitudes of the medical profession to specialization', *Bulletin of the History of Medicine* 12 (1942), 343–54

Russell, B., *St John's Hospital for Diseases of the Skin, 1863–1963.* London, 1963

Seidler, E., 'An historical survey of children's hospitals', in *The Hospital in History*, ed. L. Granshaw and R. Porter. London, 1989

Schilling, R., 'Donald Hunter, 1898–1978: editor BJIM 1944–50', *British Journal of Industrial Medicine* 50, 1 (1993), 5–6

Siena, K., *Venereal Disease, Hospitals and the Urban Poor: London's 'Foul Wards', 1600–1800.* Rochdale, NY, 2004

Sillitoe, P., 'Making links, opening out: anthropology and the British Association for the Advancement of Science', *Anthropology Today* 20, 6 (2004), 10–15

Skipp, V., *The Making of Victorian Birmingham.* Birmingham, 1996

Smith, D., *Conflict and Compromise: Class Formation in English Society, 1830–1914.* London, 1982

Smith, E., and B. Cottell, *A History of the Royal Dental Hospital of London and School of Dental Surgery, 1858–1985.* London, 1997

Smith, E., 'The Royal Dental Hospital of London: the first fifty years', *Dental Historian* 33 (1998), 19–30

Sontag, S., *Illness as Metaphor and AIDS and its Metaphors.* London, 1990

Sorsby, A., 'Nineteenth century provincial eye hospitals', *British Journal of Ophthalmology* 30 (1946), 501–46

Spencer, E. M., 'Notes on the history of dental dispensaries', *Medical History* 26 (1982), 47–66

Stammers, A. F., 'The Birmingham Dental Hospital and School', *British Dental Journal* 105, 3 (1958), 76–9

Stancliffe, F. S., *The Manchester Royal Eye Hospital, 1814–1964.* Manchester, 1964

Stanley, P., *For Fear of Pain: British Surgery, 1790–1850.* Amsterdam, 2003

Stevens, R., *Medical Practice in Modern England: The Impact of Specialisation on Medical Practice.* New Haven, CT, 1966

Stevenson, R. S., and D. Guthrie, *A History of Oto-laryngology.* Edinburgh, 1949

Sturdy, S., and R. Cooter, 'Science, scientific management, and the transformation of medicine in Britain *c.* 1870–1950', *History of Science* 36, 114 (1998), 421–66

Taylor, J., *The Architect and the Pavilion Hospital: Dialogue and Design Creativity in England, 1850–1914.* London, 1997

Thompson, F. M. L., ed., *The Cambridge Social History of Britain, 1750–1950*, vol. 3. Cambridge, 1990

Titmus, R. M., *Problems of Social Policy.* London, 1950

Tomkins, A., *The experience of Urban Poverty, 1723–82: Parish, Charity and Credit.* Manchester, 2006

Trevor-Roper, P., 'Chevalier Taylor – Ophthalmiater Royal (1703–1772)', *Documenta Ophthalmologica* 71 (1989), 113–22

Valentin, B., 'Robert Chessher (1750–1831): an English pioneer in orthopaedics', *Medical History* 2 (1958), 308–13

Waddington, K., *Charity and the London Hospitals, 1850–1898*. Woodbridge, 2000

—— 'Mayhem and medical students: image, conduct, and control in the Victorian and Edwardian London teaching hospital', *Social History of Medicine* 15, 1 (2002), 45–64

—— *Medical Education at St Bartholomew's Hospital, 1123–1995*. Woodbridge, 2003

Waddy, F. F., *A History of Northampton General Hospital*. Northampton, 1974

Walker Smith, J. A., 'Children in Hospital', in *Western Medicine: An Illustrated History*, ed. I. Loudon. Oxford, 1997, 221–31

Walvin, J., *A Child's World: A Social History of English Childhood, 1800–1914*. Harmondsworth, 1982

Warner, J. H., *The Therapeutic Perspective: Medical Practice, Knowledge, and Identity in America, 1820–1885*. Princeton, NJ, 1997

Waterhouse, R., *Children in Hospital: A Hundred Years of Child Care in Birmingham*. London, 1962

—— 'Portrait of a medical man: Dr John Ash and his Career', in *Medicine and Society in the Midlands, 1750–1950*, ed. J. Reinarz. Birmingham, 2007

Weatherall, M., *Gentlemen, Scientists and Doctors: Medicine at Cambridge, 1800–1940*. Woodbridge, 2000

Webb, K., *From County Hospital to NHS Trust: The history and archives of NHS Hospitals Services and Managemenet in York, 1740–2000*, vol. 1: *History*. York, 2002

Weisz, G., *Divide and Conquer: A Comparative History of Medical Specialization*. Oxford, 2006

Welshman, J., 'Dental health as a neglected issue in medical history: the School Dental Service in England and Wales, 1900–40', *Medical History* 42 (1998), 306–27

White, M. W., *Years of Caring: The Royal Orthopaedic Hospital*. Studley, 1997

Wildman, S., 'The development of nursing at the General Hospital, Birmingham, 1779–1919', *Nursing History* 4, 3 (1999), 20–8

Wilson, A., 'Conflict, consensus and charity: politics and the provincial voluntary hospitals in the eighteenth century', *English Historical Review* 111, 442 (1996), 599–619

—— 'The Birmingham General Hospital and its public, 1765–79', in S. Sturdy, ed., *Medicine, Health and the Public Sphere in Britain, 1600–2000*. London, 2002, 85–106

Unpublished Theses

Berry, A., 'Patronage, Funding and the Hospital Patient, *c.* 1750–1815: Three English Regional Case Studies.' DPhil diss., Oxford University, 1995

Franklin, R., 'Medical Education and the Rise of the General Practitioner, 1760–1860.' PhD diss., University of Birmingham, 1950

Lockhart, J., 'Women, Health and Hospitals in Birmingham: The Birmingham and Midland Hospital for Women, 1871–1948.' PhD diss., University of Warwick, 2008

Sewell, J. E., 'Bountiful Bodies: Spencer Wells, Lawson Tait, and the Birth of British Gynaecology.' PhD diss., Johns Hopkins University, 1990

Valier, H., 'The Politics of Scientific Medicine in Manchester, *c.* 1900–1960.' PhD diss., University of Manchester, 2002

INDEX